EAST ASIA:

THE ROAD TO RECOVERY

THE WORLD BANK
WASHINGTON D.C.

Cover design by The Magazine Group
Cover photo by Curt Carnemark/World Bank

ISBN 0-8213-4299-1

Library of Congress Cataloging-in-Publication Data has been applied for.

Contents

Abbreviations and Acronyms

AMC	Asset Management Company
APEC	Asian Pacific Economic Cooperation
ASEAN	Association of South East Asia Nations
BIBF	Bangkok International Banking Facility
BIS	Bank of International Settlements
CAMELOT	Capital, Assets, Management, Earnings, Liquidity, Operating Environment And Transparency
CD	Certificate of Deposit
CPI	Consumer Price Index
DIP	Debtor-In-Possession
East Asia 5	Thailand, Korea, Indonesia, Malaysia, The Philippines
EU	European Union
FDI	Foreign Direct Investment
FIDF	Financial Institutions Development Fund
FRA	Financial Restructuring Authority
GDP	Gross Domestic Product
GEP	Global Economic Prospects
GNP	Gross National Product
IBRA	Indonesia Bank Restructuring Agency
IMF	International Monetary Fund
KAMCO	Korean Asset Management Company
LPG	Liquefied Petroleum Gas
NBFI	Non-bank Financial Institutions
NEP	New Economic Policy
NGO	Non-governmental Organization
NIE	Newly Industrialized Economies
OECD	Organization For Economic and Cooperation Development
R&D	Research and Development
SET	Stock Exchange of Thailand
SITC	Standard Industrialized Trade Classification
TFP	Total Factor Productivity
TRIS	Thai Ratings and Information Services Co., Ltd.

Acknowledgments

This has been a team effort. The study was launched under the direction of Pieter Bottelier, Senior Advisor to the Vice President, for East Asia and completed under the guidance of Masahiro Kawai, East Asia's Chief Economist. Richard Newfarmer was the task manager, and was joined by Mona Haddad and Ilker Domaç as the study's principal authors. Stijn Claessens authored the chapter on the financial sector, based upon work done by Pedro Alba, Amar Bhattacharya, Swati Ghosh, Leonardo Hernandez, Peter Montiel, and Michael Pomerleano. Tamar Manuelyan-Atinc and Mike Walton authored the social sector chapter. Gordon Hughes, with inputs from Magda Lovei and Herman Cesar, contributed the chapter on the environment. Several members of the team provided valuable analytical papers that were incorporated into the study: Yuzuru Ozeki (macro and deputy task manager), Dipak Dasgupta and Kumiko Imai (trade), Simeon Djankov (corporate governance), Giovanni Ferri (corporate finance), Bert Hofman (corporate sector), Michael Pomerleano (corporate finance and governance). Also, Dieter Ernst, Kenichi Ohno, Takatoshi Ito, Warwick McKibbin and Will Martin, and Nomura Research Institute, Ltd., provided helpful background studies. David Bisbee provided invaluable research support. Several people contributed short analytical pieces and boxes: Natasha Beschoner, Pieter Bottelier, Craig Burnside, Elizabeth Chien, Hilary Codippily and Elizabeth C. Brouwer, Dipak Dasgupta and the DEC prospects group, Larry Lang, Wei Ding, Mary Hallward-Driemeier and David Dollar, E.C. Hwa, Lloyd Kenward, Aart Kraay, Kathie Krumm, Victoria Kwakwa, Felipe Larrain, Rolf Luders, Behdad Nowroozi, Kyle Peters, Caroline Robb, Sergio Schmukler, Richard Scobey, and Vivek Suri. Bonita Brindley provided generous editorial assistance.

We are particularly grateful to Joseph Stiglitz, the Bank's Chief Economist and peer reviewers, Amar Bhattacharya, Uri Dadush, Robert Holzmann, Danny Leipziger, Jed Shilling and John Williamson, all of whom provided thoughtful and trenchant comments at different stages of writing. Numerous other colleagues at the World Bank provided inputs and helpful critiques.

We gratefully acknowledge financial assistance from the Government of Japan through the Japanese Consultant Trust Fund and the service of the Japan Center for International Finance as a secretariat for consultants' work. Also, the study has benefited from the ADB-World Bank forthcoming study, *Managing Global Financial Integration in Asia: Emerging Lessons and Prospective Challenges* (to which some of the participants here contributed) as well as the forthcoming World Bank *Global Economics Prospects.*

Foreword

One year after it began, the economic storm in East Asia is still gathering momentum. The crisis has spread to financial markets around the world and now poses risks to the global economic expansion. Within East Asia, recession threatens to erode the remarkable achievements of East Asia's economic development. Some 370 million people were lifted out of poverty in the two decades after 1975. This is an accomplishment that in all likelihood will withstand even the gale force of this crisis, but there is no question that for tens of millions life will be much worse in the next few years. Deep recession has exposed millions of children to hunger, deprived parents of the means to support their families, and even triggered sporadic ethnic violence in some countries.

The depth of the crisis portends an enduring loss in human potential that will echo for years after this crisis has passed. Children are dropping out of school at an alarming rate. In Indonesia, for example, government officials report that enrollments are down from 78 percent to 54 percent. Economic pressures have forced countless families to split up, pushed teenage girls to enter prostitution, and put elderly poor into life-threatening privation.

For the crisis countries, it will take some time to recover the level of income they previously enjoyed. But how long? Will the region suffer a Latin American-like "lost decade" or will it begin to bounce back next year? The standard of living of a whole generation hangs on the answer to these questions.

In some respects, East Asia's downturn is unique. It has fused a currency crisis, banking crisis, and a regional financial panic into a particularly virulent strand of economic malady. To be sure, its components are well known: credit booms and asset price bubbles associated with poor financial regulation, or financial panics driven more from the herd instincts of investors responding to an isolated and random event rather than underlying fundamentals. Even the well regulated developed countries have experienced these problems. What sets East Asia apart is the harshness and magnitude of its combination of problems: a serial speculative attack on a regional group of countries, provoking massive capital outflows, simultaneous crises, and recession for a whole region.

It is obviously too soon to provide a definitive review of a drama that is still unfolding. Our objectives in this report were more modest: to take stock of progress in the region, highlight the factors shaping East Asia's future, and suggest broad directions of policies.

The main challenge is to restore broadly shared and sustainable economic growth. The report focuses on a three-pronged strategy:

- *reactivating growth* based upon structural reforms that will allow recovery to take hold sooner and make it enduring;
- *protecting the poor* during the crisis and ensuring they will share in recovery when it comes; and in
- *mobilizing capital* to help jump-start economic growth.

The hard work of implementation lies ahead.

The World Bank's Support

In each country of the region, the World Bank is working in tandem with governments to realize this three-part strategy. The World Bank has pledged nearly $18 billion to the East Asia crisis countries and disbursed over $8 billion in loans in the year since July 1997.

Reactivating growth on the basis of structural reforms is a high priority. The World Bank is helping governments to increase their spending in efficient ways, especially social spending. It has approved 45 major loans to the region in the first year after the crisis. The World Bank has thus helped to finance a more expansionary fiscal position, and the resulting demand will in turn help to create jobs and income. Our focus has not only been on the amount of spending but, even more important, on the *quality of spending*. In preparing and supervising these loans, the World Bank has provided policy advice and technical assistance, a dialogue that is supplemented with a steady stream of economic studies, public expenditure reviews, and conferences involving partners from the private sector and Non-governmental organizations (NGOs). Through its multi-billion dollar structural adjustment programs in Thailand, Indonesia, the Republic of Korea, and the Philippines, for example, the World Bank is helping governments to improve financial sector regulation and

supervision while it also helps these governments restructure their banking and corporate sectors. This also means improving corporate and financial disclosure, better management of debt and contingent liabilities, and implementing legal and regulatory reform. Through its project lending, especially to infrastructure, the World Bank is intensifying dialogue to instill environmental safeguards that address the environmental and natural resource problems exacerbated by the crisis. These efforts contribute to the restoration of growth, and growth that can be sustained.

To *protect the poor*, unemployed, and elderly from the social impacts of the crisis, the World Bank has supported basic health, education, targeted food subsidies, and labor-intensive and employment-generating public works. Social fund projects and stay-in-school programs have been introduced in Indonesia and Thailand, and improvements to social safety nets (including labor market, pension reform, delivery of social services, poverty targeting) have been developed in other countries. Over the long-term, the World Bank is working toward improving the social and human sustainability of growth to address the social shortcomings of East Asian development—growing inequality and lack of formal social safety nets such as healthcare and unemployment insurance—while protecting and reinforcing the region's social successes—education, health, and quality of life improvements.

The Bank is redoubling its efforts *to mobilize external resources* for the region. Arguably the World Bank's most important contribution is not in the capital it provides but in helping the region regain the confidence of domestic and foreign investors through sound policies. Restoring confidence in the future is the secret to attracting new capital inflows. Often there is a lag between the time sound policies are adopted and the return of market confidence. The World Bank intends to play a leadership role in mobilizing capital during this period. It will increase its own lending up to the limits of its own prudential regulations as long as the pace of domestic policy reforms warrants. Beyond this, in conjunction with other partners—the International Monetary Fund, the Asian Development Bank, governments, and the private sector—the World Bank will be looking for new ways to mobilize capital to help jump-start growth.

The task ahead is enormous. The crisis is as important to East Asia as the debt crisis of the 1980s was to Latin America. As that crisis changed irreversibly the economic and political institutions of the day, so too throughout East Asia societies are changing dramatically in ways that none would have predicted only 18 months ago. Virtually all of the countries in East Asia are transforming the old ways of conducting their business and politics. Companies that borrowed freely and frequently using only the collateral of unfolding rapid growth are being subjected to a new discipline. Banks that borrowed in yen or dollars and loaned in local currency using a nod from government as their only hedge, are being subjected to greater supervision. Enterprises and banks are undergoing ownership and organizational changes as profound as those in Latin America during the 1980s or even in the United States during the 1930s. Though it is too soon to say with certainty, companies and banks may well emerge with less concentrated ownership, greater representation and transparency for minority shareholders, including foreigners, and greater discipline from competition in both product and capital markets. Similarly, governments and public governance are changing in historic ways. Even as they shoulder burdens from past implicit guarantees to the private sector, governments are reorganizing themselves to reduce these contingent liabilities and their direct role in resource allocation. They also are assuming new responsibilities. As traditional rural family ties breakdown under pressures of urbanization, societies everywhere in the region are looking to governments for help in ensuring the welfare of the poor, the unemployed, the sick, and the elderly. In the backdrop, a new politics of governance—from Korea in the North to Indonesia in the South—seems to augur a new openness, concern for corruption, and accountability. The journey to recovery, filled with uncertainty to be sure, is set on a historic course that will shape the future of East Asia's children.

Jean-Michel Severino
Vice President
East Asia and the Pacific Region
World Bank

Executive Summary

East Asia's financial crisis quickly has deteriorated into an economic and social crisis. Real wages have plummeted, and the region's major cities are filled with idled workers looking for jobs. In the countryside, the combination of drought-parched lands and dried-up rural credit has threatened the livelihood of many. Since this comes after three decades of rapid growth, a whole generation of workers and farmers has never known these hardships, and societies have developed few formal mechanisms to ease their plight.

This study presents an analysis of the crisis, provides a report card on progress within the region, and suggests policy directions that will affect the pace of recovery. *The most urgent task ahead is to restore the conditions for robust economic growth throughout the region.* This is particularly true for Thailand, the Republic of Korea, Indonesia, and Malaysia, where recession has been unrelenting and severe. The other, smaller developing countries in East Asia are feeling the ripples of the crisis, and are fighting off deep recession. The economies of Taiwan (China), Vietnam, and China have so far avoided the recession, but they too have been pushed below their trend-line growth paths.

Origins of the crisis

Even as growth was improving the livelihoods of the poor, it had begun creating several sources of vulnerability in the mid-1990s. The region's very success—rapid growth, conservative economic management and low indebtedness—made it attractive to private capital. These inflows, while spurring growth, were intermediated through poorly regulated domestic financial systems and helped fuel domestic credit expansion. The pace and pattern of growth, interacting with often-undisciplined capital inflows, produced three weaknesses in the foundation of East Asia's growth:

- Large *current account deficits*, financed with short-term flows, exposed East Asian economies to sudden reversals.

- Liberalization of domestic *financial markets without adequate prudential regulation* and supervision allowed banks and corporations to assume unhedged foreign borrowing positions that left them vulnerable to sudden currency fluctuations.
- *Companies*, in the absence of fully developed bond and equity markets, borrowed heavily from banks to finance their rapid expansion, and in the process became very highly leveraged. This left them vulnerable to interest rate surges.

When markets became worried about the sustainability of the fixed exchange rate in Thailand, capital inflows became outflows. Asset values plummeted—particularly equities and property—and suddenly turned what had been a virtuous circle into a vicious one. Falling asset values reduced wealth and imposed balance sheet losses on financial agents, demand fell, and contracting markets produced greater outflows. Finance stampeded to safe havens, making the situation worse.

Main challenge: Restoring growth

The main challenge today is restoring broadly shared and sustainable growth for the region. Three elements form the basis of a strategy.

Enacting structural reforms to restore high quality economic growth. The only way to reverse the income losses imposed upon the poor is for countries to reactivate economic expansion. But the quality of growth matters. If it is not environmentally sustainable, leaves out the poor, or is cut short because of inadequate structural foundations, recovery will not achieve its promise. A pre-requisite is reactivating demand. Exports are growing slowly because neighboring countries are also in recession, investment is hobbled by systemic insolvency in the banking and corporate sector, and declining incomes and wealth have depressed consumption. In Indonesia, Korea, and Thailand, 20–65 percent of firms are estimated to have balance sheet losses greater than equity. Insolvent, highly leveraged companies cannot service their debt. Non-performing loans in those countries are thus estimated to range from 20–40 percent. The situation has created a self-re-enforcing downward spiral: recession forces corporations to delay or default on bank payments, and, as the amount of non-performing loans rises, banks' cash flows are squeezed, forcing them to contract new lending to illiquid corporations and call-in even good loans to raise cash, further deepening the recession. This report therefore focuses on structural reforms that will reactivate demand in a sustainable fashion: speeding up the process of financial and corporate restructuring, establishing a better framework for financial and corporate governance, enhancing public sector management, and improving environmental policy. Only by progressing on this combined agenda can countries ensure that growth will be sustainable and of high quality.

Second, *ensuring that low-income groups are protected during crisis and then share in eventual recovery.* If output were to fall by a cumulative 10 percent over the next three years and income distribution worsens by 10 percent, the number of poor people in Indonesia, Thailand, Malaysia and the Philippines would more than double—from some 40 million to more than 90 million. This is an unlikely but still possible scenario, and underscores the importance of renewing growth. The report lays out an agenda of pro-poor fiscal policies, suggests ways to maintain incomes of the poor, and focuses on enhancing social services that cushion the worst effects of recession upon the poor. Reforms in pension systems, labor markets, and education can help incorporate low-income groups into a sustained economic expansion.

Finally, the international community must do what it can to restore *international capital flows.* The region has suffered a massive swing in private capital. Domestic policies that inspire investor confidence are a necessary condition for renewing private capital inflows. With policies in place, a concerted effort to mobilize additional finance would mitigate the pressure on consumption levels in the region and spur growth. If an additional $10 billion in external finance could be mobilized and it were used to finance an additional fiscal stimulus, it would provide a strong impetus to growth. If some of the spending were focused on low income groups, it could mitigate the worst effects of the crisis. The report does not delve into specific mechanisms for mobilizing finance, a discussion that is transpiring in international fora and elsewhere, but it is nonetheless essential that this challenge be faced squarely.

Looking ahead

Recovery is likely to take longer in East Asia than in Mexico and Argentina in 1994-95 because of the problem of corporate and bank insolvency and because of the regional scope of recession, including Japan. The global economy has been so far supportive, but events in Russia and in global financial markets in recent months raise worrisome signs that even that bright spot may be dimming. The recent floods in China also weigh on the region's prospects. An expanding global economy is arguably the most important element in East Asia's recovery.

It would be easy, however, to be overly pessimistic about the region's future. Countries throughout the region are moving swiftly to enact new policies and adopt new, more transparent ways of doing business. They have shown themselves willing to work extraordinarily hard and sacrifice today for benefits tomorrow. Witness the region's continuing high savings rates. If the pace of reform accelerates and if the international community responds positively, the region will soon find itself on the road to recovery.

chapter one

East Asian Crisis: An Overview

In mid-October, 21 year old Sugiyanto was still swinging a shovel at a Jakarta construction site. Six months earlier he had made a personal pilgrimage by overnight bus to the capital from the village of Banjarjo in central Java. "On TV, it looked so easy to make money in Jakarta." The money, however, has since dried up. Indonesia's financial crisis has brought many construction projects to a halt. With new jobs scarce, Sugiyanto slunk home to Banjarjo in early November—only to find that his father's rice paddies had dried up too. Months of dry weather have turned the fields in to a parched brown expanse. No work, no monsoon, no escape. For Sugiyanto there's nothing to do all day but slump over a motorcycle, hoping to cadge a few faded bills in return for offering lifts. There are few takers. Villagers would rather spend their money on water.—Margot Cohen, "Unlucky Country," *Far Eastern Economic Review*, December 25, 1997.

After three decades of remarkable expansion, the economies of East Asia have gone into a tailspin. The once booming economies of Thailand, the Republic of Korea, Indonesia, and Malaysia will contract this year. Singapore, Hong Kong (China) and Taiwan (China), with their strong financial systems and high reserves have, so far, managed to fight off the worst of the contagion, but have seen their export markets and businesses contract. The transition economies, partially protected with their semi-closed capital accounts and low ratios of short-term debt to reserves, have emerged with the added challenges of diminished prospects for exports and capital inflows.

The smaller economies, from Mongolia to Fiji, are buffeted by the storm around them. The Solomon Islands may contract by 10 percent or more in 1998.

The currency and financial crisis has quickly deteriorated into a social crisis. In the past, steady economic expansion provided the underpinnings to the livelihood of the poor, and substituted for a formal social safety net. Today that is gone. Unemployment is rising. Real wages of low-income urban workers have plummeted, and the region's major cities are filled with idled workers looking for ways to make a living. Inflation has risen, with the possible effects of worsening income distribution and further reducing the real wages of low-income groups. To make matters worse, drought has parched much of the region's otherwise fertile land, making it difficult for farmers to take advantage of higher food prices. The effects of falling incomes are felt most severely by poor women and children. Also, in some countries economic pressures have ignited latent social prejudices against minorities and immigrants.

The effects go beyond the poor. The currency fall and crash of the equity markets has wiped out savings of the middle class and newly rich. The decline in equity values in the region has surpassed US$400 billion since July 2, 1997. Meanwhile, efforts to improve the quality of life of all East Asian citizens through greater social and environmental investments have stalled.

Signs of a new, if fragile, financial stability are appearing in four of the five crisis-affected Asian countries (Korea, Philippines, Malaysia, and Thailand). The Philippine economy has thus far come through with surprising vitality. Thailand and Korea, after suffering collapse in their financial systems, have established firmer values for their currencies and are rebuilding reserves. Indonesia is still fighting to regain a modicum of stability. However, the economic recovery that all had hoped would come soon is not yet in sight. While the prospects are uncertain, it is clear that the changes wrought by the events of 1997 will be as profound as those brought on by the debt crisis in Latin America during the 1980s. Although unemployment in East Asia will probably not reach the levels of some countries in Latin America, a far greater share of the population in East Asia is living just above the poverty line, so any substantial slowdown puts their livelihood at risk. Furthermore, the shock of recession to the middle class, coming as it did after the most rapid sustained expansion in human history, will undoubtedly be as profound. The end of the 20th century for East Asia is changing the way business is conducted, the way resources are allocated, and the very economic and, in some cases, political governance of countries.

This study looks at these changes and focuses on policies for a sustainable recovery. It is designed to be a snapshot of where the region stands, a progress report on the enormous changes that have been made in the last year, and an analysis of the remaining obstacles to establishing a firm economic recovery. But, the region cannot be satisfied with a short-lived growth spurt. It must aspire to nothing less than recapturing the lost growth momentum of the last three decades. Subsequent chapters look at those policies. This chapter reviews the gains of the past and the causes of the crisis.

Was the miracle real?

East Asia's achievement of spectacular welfare gains in the last two decades is beyond dispute. Poverty has declined, not only in breadth (the number of poor) but also in depth (severity of poverty). Life expectancy at birth, infant mortality rates, and literacy indicators have all improved in tandem, generating real improvements in peoples' lives. The region succeeded in converting persistently high growth rates into improvements in welfare because growth, supported with widespread social services, created jobs for the poor and enormous opportunities to expand productivity. The miracle was real and tangible.

The region reduced the number of people living in poverty by half in the last 20 years. As figure 1.1 shows, the number of poor living below the international poverty line of US$1-a-day[1] was reduced from 720 million to 350 million. Moreover, the rate of decline accelerated over the past decade: the total number of people in poverty fell by 27 percent during the period 1975–85, and fell an additional 34 percent during the period 1985–95. This pace of poverty reduction was faster than in any other region of the developing world, and, as a result, the share of the world's poor living in East Asia has declined. While six out of ten East Asians lived in absolute poverty in 1975, roughly two in ten did in 1995.

Within East Asia neither poverty levels nor rates of decline were identical across countries. In 1975, China and Indonesia alone accounted for 92 percent of the region's poor, largely because they were the two most populous countries. Since 1975, however, both countries have recorded substantial declines in poverty, 82 percent in Indonesia and 63 percent in China. In absolute terms, the number of poor decreased by more than one-half in China and fell by almost three-fourths in Indonesia (the head count declined from 64.3 percent in 1975 to 11.4 percent in 1995). By 1995, the two countries accounted for 84 percent of the region's poor. Although Indonesia's record was remarkable, Malaysia had the largest proportional reduction between 1975 and 1995 (95 percent decline, from 17.4 to less than 1 percent) and Thailand was a close second (90 percent decline from 8.1 percent to less than 1 percent).

Propelling these achievements was a high performance engine of economic growth. Several factors lay behind this growth performance. Governments generally:

- Kept inflation low and exchange rates competitive through conservative macroeconomic policies
- Invested in human capital through public expenditures on education

- Encouraged high rates of savings by keeping interest rates positive in real terms and by effectively protecting deposits in financial institutions[2]
- Limited price distortions
- Encouraged absorption of foreign technology
- Avoided implicit taxation and other biases against agriculture.

Several studies have confirmed that high rates of savings, investment in human capital, and stable macroeconomic policies are key determinants of growth.[3]

Why did East Asia falter?

Such remarkable economic and social performance made the sudden downward spiral of the East Asian 5 all the more startling. Several structural problems were well known and analyzed prior to the collapse of the Thai baht in July 1997. Did these structural problems finally produce the exhaustion of the East Asian model, much as import substitution in Latin America became fully exhausted during the crisis decade of the 1980s? Or, was the crisis in East Asia the result of short-term macroeconomic mistakes and financial panic, a type of macro-financial accident?

Rapid growth, urbanization, and industrialization were spawning new and difficult development problems prior to the crisis. These were building in three dimensions. First, rapid growth, in the absence of sophisticated financial and capital markets and with a large government presence, left the corporate and financial sectors unusually reliant on financing long-term investment with short-term debt capital (this will be discussed below). Second, economic growth was undermining the traditional protection mechanisms for the unemployed, the sick, and the elderly. East Asia relied on high personal savings and family ties to provide security for its elderly. It came to rely on growth itself to provide an ever more buoyant labor market. The forces of growth, with their demands for an increasingly mobile labor force, migration, and wider scope for personal consumption, were putting strains on traditional ways of solving social problems. In the transition countries of China and Vietnam, the old commune and state enterprise system of welfare was under analogous strains with the spread of markets. In the wealthiest countries, lifetime employment guarantees in the corporate sector were proving increasingly out of tune with

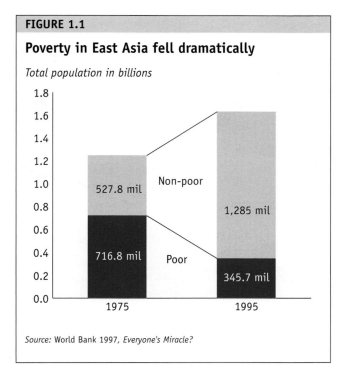

FIGURE 1.1

Poverty in East Asia fell dramatically

Total population in billions

Source: World Bank 1997, *Everyone's Miracle?*

the modern economy's needs for rapid change and flexibility. Third, a weakness of a different kind resulted from the exploitation of national resources, particularly forests. Southeast Asian growth was fueled, in part, by over-logging, intensive exploitation of fisheries, and wasteful agricultural practices. Although national income accounts are difficult to adjust for environmental damage, some estimates are that Malaysia's growth in gross domestic product (GDP) would have been approximately 20 percent less if adequate allowance had been made for resource depletion.

Nonetheless, there is not much evidence that these long-term development problems alone were enough to drag down growth, much less precipitate a sudden reversal of fortune. Productivity growth was generally within the normal range for developing countries.[3] In that sense the "miracle" was no miracle at all. Rapid growth relative to other countries was achieved by dint of sacrifice reflected in East Asia's famously high savings rates, hard work as reflected in the dramatic increases in labor force participation rates, and investments in education as reflected in the skill level of the work force. Productivity, per se, is less important than increases in per capita income, whatever the source, and East Asia simply out-performed other regions of the world by this more meaningful measure.

Even if declining returns to investment eventually were to set in, the question is when and whether they would be sufficient to precipitate sharp slowdown or crisis. Comparisons with a conventional Solow model suggest East Asian performance exceeded predictions in most countries (see box 1.1). Moreover, any growth slowdown associated with diminishing returns is likely to be well into the 21st century, not the mid-1990s. The main sources of the crisis will have to be found elsewhere.

Emergence of structural vulnerability

Three forces interacted to leave some countries in the region—notably Thailand, Korea, and Indonesia—vulnerable to external shocks: a burgeoning availability of private capital, especially short-term capital, that was in search of higher returns; macroeconomic policies that permitted capital inflows to fuel a credit boom; and newly liberalized, but insufficiently regulated

financial markets that were growing rapidly. The scenario played out as follows: The push from global capital markets, often without due diligence and beyond prudent limits, interacted with poorly regulated domestic financial systems to fuel a domestic credit expansion. This manifested itself as an asset price bubble, particularly in Thailand, and added to the excessive debt of already over-leveraged firms, which exposed the region to the shocks of changing investor expectations.

Ready availability of capital

Globalization of financial markets has been occurring at a dizzying pace. From 1990 to 1997, the volume of private capital flows to developing countries rose more than fivefold—from US$42 billion in 1990 to US$256 billion in 1997. While world trade grew by about 5 percent annually, private capital flows grew by nearly 30 percent annually. The most mobile forms of flows, commercial bank debt and portfolio investments, set the pace.

Propelling this expansion was an aggressive search for ever higher returns to capital. "Emerging markets" were booming, and offered greater profitability than investments in developed countries. Banks and financial institutions, often trapped in slow-growing but

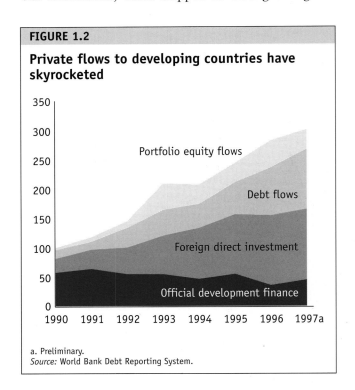

FIGURE 1.2

Private flows to developing countries have skyrocketed

Portfolio equity flows

Debt flows

Foreign direct investment

Official development finance

1990 1991 1992 1993 1994 1995 1996 1997a

a. Preliminary.
Source: World Bank Debt Reporting System.

BOX 1.1

Of paper tigers and productivity growth

In an influential 1994 article, Paul Krugman drew attention to the then-novel findings of a number of economists that growth in East Asia had been due more to increases in inputs rather than increases in the efficiency with which those inputs were used. This finding prompted him to refer to the economies of East Asia as a collection of "paper tigers" whose growth rates were bound to decline with the onset of diminishing returns. Have events vindicated this view?

Rapid factor accumulation unaccompanied by rapid productivity growth logically implies that growth rates must eventually decline as diminishing returns set in. The key question, of course, is when this will occur, and by how much. To provide a back-of-the-envelope answer to this question, the figure below plots the log-level of per capita income, and the log-level of income predicted by the Solow growth model, assuming a measured total factor productivity (TFP) growth rate of 1 percent per year, for Indonesia and Korea. In addition, the model is used to project forward income levels, assuming unchanged TFP growth and savings rates and labor force growth rates that are equal to their averages over the period 1991-95. The figure suggests that, consistent with the conventional wisdom, a simple model of growth driven primarily by factor accumulation and with only modest productivity growth provides a reasonably good description of these countries' historical growth experiences.

The figure plots the log of actual GDP per worker and that predicted by the Solow model, assuming a constant TFP growth rate of 1 percent per year. The figure also shows that, consistent with the theory, growth rates can be expected to decline over time as diminishing returns set in. However, the decline in growth rates is fairly gradual. As shown in the table for these and for the other three crisis economies, the onset of diminishing returns can be expected to account for only rather moderate declines in growth rates over the next few years relative to historical averages.

What would Solow say?

(Log of GDP per worker)

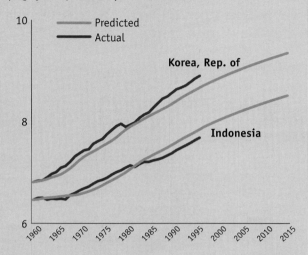

What do these calculations imply for recent events in East Asia? The financial crisis has undoubtedly dampened growth prospects for these countries for the next several years. However, the decline in growth rates is much larger and more dramatic than what would be expected as a result of the simple petering out of rapid growth due to the onset of diminishing returns. Although this is surely to be an important feature over the longer term, it has little relevance for the recent growth collapse, a point Mr. Krugman himself has made.

Diminishing returns and other culprits
(Average annual growth in GDP per worker, percent)

	1960–96		1997–2000		Difference	
	Actual	Predicted by Solow Model	GEP projections	Predicted by Solow Model	GEP	Solow
Indonesia	3.5	4.1	-0.1	4.1	-3.6	0.0
Korea, Rep. of	5.9	5.5	0.4	4.3	-5.5	-1.2
Malaysia	3.7	4.4	1.5	4.2	-2.2	-0.2
Philippines	0.7	2.9	1.2	2.1	0.5	-0.8
Thailand	4.0	4.1	-1.8	3.9	-5.8	-0.2

GEP projections taken from the February 1998 Global Economic Prospects Update.

Source: World Bank staff.

highly competitive home markets, scanned the globe for investment opportunities.

The very success of East Asia made it an ideal location and the combination of rapid growth, low debt ratios, and a history of sound macroeconomic management attracted capital like a magnet.

From inflows to credit boom: Macroeconomic and exchange rate policy

Macroeconomic policy in East Asia inadvertently created incentives for private agents to take advantage of the easy access to international capital, and these flows financed a domestic credit boom. Indonesia, Korea, Malaysia, and Thailand experienced a sharp acceleration of domestic demand. The macroeconomic policy mix used to deal with the overheating pressures and capital inflows in the 1990s added an impetus for further inflows, particularly for the accumulation of short-term, unhedged external liabilities. However, tightening monetary policy in an effort to sterilize inflows and curtail credit expansion increased domestic interest rates, as well as the differential between domestic and foreign rates. This had the perverse effect of creating further incentives for investors to borrow abroad to make local investments. On the fiscal side, governments throughout the region had generally run fiscal surpluses from the late 1980s, and were unaccustomed to using fiscal policy as a macroeconomic instrument. While fiscal policy in most East Asian countries remained conservative in a medium-term structural sense, the fiscal impulse (the change in the fiscal stance) turned positive at the time when these economies were experiencing demand pressures.[5]

Most of the Association of Southeast Asian Nations (ASEAN) countries adopted a nominal anchor policy by pegging loosely to the U.S. dollar in the run-up to the crisis, switching from real exchange rate targeting in the earlier period. Informal pegs to the U.S. dollar encouraged capital inflows due to large interest rate differentials. Predictable nominal rates encouraged unhedged external borrowing. A wedge was driven between the actual and equilibrium real exchange rates due to a loss of competitiveness and declining corporate profitability on the one hand, and a real appreciation on the other. Exceptions were Singapore and Hong Kong (China) where labor markets were flexible and productivity

gains were high. Thus, the link to the U.S. dollar in Hong Kong (China) or the strong currency policy in Singapore did not result in a real exchange rate misalignment. To further complicate matters, the yen depreciated against the U.S. dollar throughout much of 1996, so the pegged currencies lost competitiveness against the important yen market. But, the most important effect was the incentives the policy gave to borrow abroad. Exchange rate policies played a particularly large role in motivating capital flows. By reducing the perceptions of exchange rate risks, incentives to hedge external borrowing were suppressed and, moreover, the relatively narrow exchange rate movements created a bias toward short-term borrowing.

Between 1994 and 1997, the net private capital inflows as a share of the rapidly expanding GDP increased throughout the East Asia 5. The exception was Thailand where, by 1994, the net private capital inflows had already reached 14.5 percent of GDP (see figure 1.3).

East Asia generally absorbed nearly 60 percent of all short-term capital flows to developing countries. In the mid-1990s, much of the short-term private capital came from Japanese banks as they followed their corporate foreign investors into Korea and Southeast Asia. The Europeans soon followed in an aggressive search for

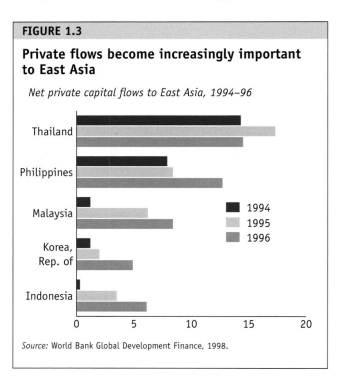

FIGURE 1.3

Private flows become increasingly important to East Asia

Net private capital flows to East Asia, 1994–96

Source: World Bank Global Development Finance, 1998.

profits. By 1996, the Bank for International Settlements (BIS) reported that European Union (EU) banks' outstanding bank loans amounting to US$318 billion; the Japanese banks had US$261 billion; and the U.S. banks had US$46 billion (WEO, 7).

These inflows fueled the domestic credit boom throughout most of the region. In the East Asia 5, broad money (M2) expanded at a near 20 percent annual rate in 1996 and 1997. This was nearly twice the rate of China, Taiwan (China), Hong Kong (China), and Singapore—countries that would later fare better in the storm of speculative attacks. The credit boom, in turn, led to an increase in assets prices, creating the appearance of high returns. Property values in Bangkok, Seoul, and Jakarta rose at double digit rates through 1996. Rising asset prices provided greater collateral to banks, and led to greater lending. At the same time, middle- and upper-class owners of these assets, feeling more well-heeled, consumed more freely. Rising aggregate demand encouraged yet more foreign borrowing.

Weak financial systems led to poor investments and excessive risks

As capital inflows increased, the quality of intermediation became increasingly important. Invested in high-return activities to creditworthy borrowers, these capital inflows had the potential to spur East Asian growth. However, incremental additions to investments appear to have yielded a lower return. As indicated in figure 1.4, the incremental capital-output ratio in Thailand and Korea, after some fluctuations in previous decades, rose every year after 1988.

East Asian countries receiving foreign capital primarily through the domestic banking system or through direct corporate borrowing became more vulnerable than countries relying predominantly on foreign direct investment.[6] This was especially true in Thailand. Private decisions that resulted in an excessive buildup of risky forms of leverage on the balance sheets of financial institutions and non-financial corporations, in particular of short-term foreign currency debt in excess of foreign currency resources available on short notice. In several East Asian countries in the late 1980s, short-

term debt relative to overall external liabilities began rising sharply.

Capital inflows and the credit boom increased vulnerability in two dimensions. On the one hand, the ratio of short-term debt to foreign reserves, a rough measure of a country's ability to meet its current obligations from its own liquid resources, rose sharply from 1994 to 1997, except for Indonesia, where it remained at high levels. In the three most-affected countries—Korea, Indonesia, Thailand—short-term debt-to-

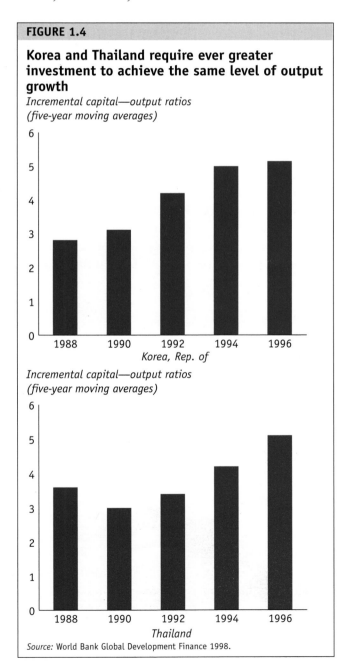

FIGURE 1.4

Korea and Thailand require ever greater investment to achieve the same level of output growth

Incremental capital—output ratios (five-year moving averages)

Korea, Rep. of

Incremental capital—output ratios (five-year moving averages)

Thailand

Source: World Bank Global Development Finance 1998.

reserves ratios had risen to well over 150 percent by June 1997. Malaysia and the Philippines were not as badly exposed, with ratios at less than 100 percent. Credit growth was evident in the high ratio of broad money to reserves, and the two were correlated, as seen in figure 1.5. A broader measure of vulnerability, the ratio of M2 money to reserves, indicates the potential for a "run" on the foreign exchange reserves of a country with a fixed exchange rate by its own residents when there is a loss of confidence in the local currency. Countries with exchange controls and less open capital accounts are less vulnerable than this measure would otherwise indicate because of the difficulty in shifting funds out of the country.

Patterns of indebtedness varied across countries. In Thailand, finance companies and banks, availing themselves of extremely low-interest, yen-denominated loans, borrowed through government sanctioned channels to invest in real estate. Financial institutions' net foreign liabilities rose from 6 percent of domestic deposit liabilities in 1990 to one-third by 1996 (Global Economic Prospects (GEP), 1998). Korean banks also increased their exposure to foreign borrowing. In Indonesia, however, corporations became the primary borrowers from foreign sources, with much of it coming from "off-shore."

Three microeconomic factors accentuated the incentives to borrow abroad. First, the implicit insurance— for example, the fixed exchange rate—provided to financial institutions motivated excessive risk-taking, including large foreign exchange risks, that were passed on to the rest of the domestic economy. Second, high domestic funding costs and market segmentation added to the incentives to borrow abroad. In Thailand during the period 1991–96, domestic financial intermediation costs accounted for 28 percent of the nominal baht interest cost.[7] The domestic cost of funds was significantly higher than the costs of borrowing "off-shore," even after taking into account exchange rate risks, which only added further incentive to borrow foreign funds. Since this access to foreign markets was only available to the largest and best credit corporations, these firms and banks enjoyed a market advantage, and undoubtedly used their access to political leaders to protect their position, making it more difficult for regulators to limit "off-shore" borrowing to prudent levels. Third, the creation of "off-shore" financial markets in which local corporations could, because of regulatory and tax advantages, obtain lower-cost finance than in domestic markets. This situation was the most severe in Thailand.

The inflows also fed into a system of corporate finance that heightened risks from abrupt changes in interest or foreign exchange rates. The corporate sector had grown rapidly during the previous decades in a context of under-developed bond markets and over-reliance on bank financing. The debt-equity ratio of Korean corporates, for example, was over 317 percent by the end of 1996, twice the U.S. ratio, and four times the Taiwanese ratio. The top 30 Korean chaebols had even higher leverage, on average more than 400 percent in 1996. Correspondingly, interest burdens are very high in East Asian countries. In Korea, for example, the interest-expenses-to-sales ratio of all manufacturing corporations in 1995 was about 6 percent, compared to 2 percent for Taiwan (China) and 1 percent for Japan. This would present a painful dilemma to macroeconomic policy makers when the crisis hit: they could use interest rate adjustments to maintain exchange rate stability but only at the cost of imperiling their highly leveraged corporate sectors and creating a domestic liquidity crunch.

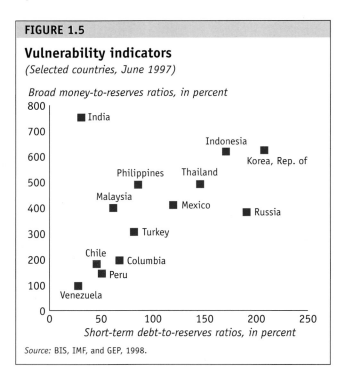

FIGURE 1.5

Vulnerability indicators
(Selected countries, June 1997)

Broad money-to-reserves ratios, in percent

Short-term debt-to-reserves ratios, in percent

Source: BIS, IMF, and GEP, 1998.

In retrospect, it is also clear that the regulations necessary to manage the integration of global external finance had not kept pace with capital inflows. Inconsistent reforms and inappropriate sequencing of liberalization added to the buildups of vulnerabilities. For example, licensing and supervision regulation of merchant banks in Korea permitted groups of companies to own both banks and the same groups of firms to whom they were lending. In Indonesia, the number of banks expanded very rapidly and the supervisory authorities spent too little time screening the integrity of owners and managers to keep out applicants with poor prospects or fraudulent ventures. In Thailand, the scope of finance companies' activities greatly increased in the 1990s without a commensurate improvement in their supervision. In several East Asian countries, the capital account was liberalized for inward and outward flows for foreign investors; domestic investors, however, did not always have the opportunity to invest abroad and thus, could not diversify their risks. Finally, throughout the region, regulations requiring prudential management of currency risks, credit evaluation, and public financial reporting were wholly inadequate.

The time bomb was loaded. Rising global liquidity fed huge amounts of capital into a poorly regulated institutional setting with limited transparency, and related party lending, often with negligible due diligence from foreign lenders. Implicit and explicit government guarantees on the exchange rate and selected investments fed into a domestic credit boom that macroeconomic policy failed to manage. East Asian countries had taken risks that left them exposed to shocks in several ways:

- Widening current account deficits, financed with short-term debt, exposed the economies to sudden reversals in capital inflows.
- Weaknesses in the under-regulated financial sector had allowed expansion of lending into risky investments of inflated values, often with currency and maturity mismatches, which exposed entities to exchange rate risks.
- Corporations, often with insider relationships with banks and having little incentive to use capital efficiently, became even more highly leveraged when presented with additional funding options from abroad, which exposed them to relatively small interest rate shocks.

This created a potentially explosive situation that only required detonation.

Trigger

Macroeconomic imbalances and financial sector weaknesses were most pronounced in the case of Thailand: the current account deficit, which reached very high levels of 8 percent of GDP, was financed by short-term inflows. The heavy inflows and credit boom channeled substantial investment into real property, creating an asset price bubble. The private sector had borrowed huge amounts from abroad and, taking advantage of the promise of a pegged exchange rate, did not hedge against foreign currency risks. Thai borrowers, many of which were under-regulated finance companies, invested in the booming property market. In the mid-1990s, an investor could borrow in yen at near zero interest rates and invest in Bangkok skyscrapers, whose expected annual return was 20 percent.

In 1996, export growth hit a wall. After growing 20 percent in 1995, exports actually contracted by 1 percent in 1996. Although all East Asian exports had slowed in conjunction with diminished world demand, Thailand was the worst hit. The impact was the result of three elements: the loss of wage competitiveness associated with appreciation; the demand for its products, particularly electronics, slumped badly in world markets; and because growth in its markets, notably Japan, slowed sharply. At the same time, prices of real assets stopped growing. Vacancy rates increased in 1996 as the supply of office space began to outpace demand. The finance companies began to experience serious difficulties in early 1997. The government response to furnish them with liquidity, only added to the supply of funds in the market ready to attack the peg.

Equity investors were the first to withdraw. The stock market peaked for the year in February, and fell by more than 30 percent by year's end. As the yield curve tilted against Thai borrowers, short-term borrowing became increasingly common. Perceptions began to take hold in the market that asset prices were getting too high and the exchange rate was misaligned. In early 1997, total private capital flows started to taper off. In the first half of 1997, bond issues and syndicated loans fell by 30 percent relative to the same period in the pre-

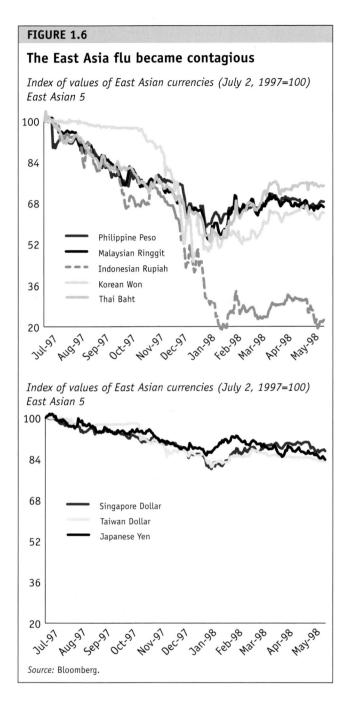

FIGURE 1.6

The East Asia flu became contagious

Index of values of East Asian currencies (July 2, 1997=100)
East Asian 5

Philippine Peso
Malaysian Ringgit
Indonesian Rupiah
Korean Won
Thai Baht

Index of values of East Asian currencies (July 2, 1997=100)
East Asian 5

Singapore Dollar
Taiwan Dollar
Japanese Yen

Source: Bloomberg.

large volume of liquidity to support them. Even more baht were chasing fewer dollars. Soon capital began to seek safe haven, and reserves fell. Finally, on July 2, 1997, the government yielded to the market forces and abandoned the peg. The crisis that was to rock East Asia and reverberate throughout world financial markets had begun.

Contagion

The Thai devaluation triggered a withdrawal of capital from the region as a financial panic progressively set in. Developments in Thailand caused investors to look more critically at weaknesses they had previously ignored. In the process, they discovered new information that amplified their concerns, especially about the health of the financial system and the magnitude of short-term debt. Market doubts were compounded by the lack of transparency about the financial and corporate sectors, and thus, about the magnitude of contingent liabilities. Once investors lost confidence that reserves would cover short-term debt, both foreign and domestic investors scrambled to get out.[8] Markets became much less forgiving. The lack of a mechanism for orderly workouts of corporate and bank debt undoubtedly contributed to the full-scale financial panic that swept Thailand, Korea, and Indonesia, and to a lesser extent Malaysia.

Contagion produced simultaneous declines in asset prices and spurred capital outflows.[9] Within the space of six months, capital outflows from the region erased the inflows of the first semester, and turned the net flow to a negative US$12 billion. As shown in table 1.1, in the space of a year net capital flows reversed by more than US$100 billion.

TABLE 1.1
Private capital flows reverse...with a vengeance

Net private capital flows in five East Asian economies (US$ billion)	1996	1997
Private flows (net)	97.1	-11.9
Non-debt flows	18.7	2.1
Foreign direct investment	6.3	6.4
Portfolio equity investment	12.4	-4.3
Debt flows	78.4	-14.0
Banks	55.7	-26.9
Non-banks	22.7	12.9

Source: IIF, Capital Flows to Emerging Market Economies.

vious year. Confidence took a further hit when Somprasong Land defaulted on a Eurobond issue.

When the baht came under attack in February 1997, the government intervened heavily to support the peg. The central bank issued some US$23 billion in forward foreign exchange contracts at a time when reserves were hovering around US$25 billion. As investors' perceptions continued to sour, the finance companies came under pressure, and the government had to pump a

Competitive devaluations: An unlikely cause for contagion

If the East Asian countries compete in the same export markets, a devaluation of one currency places competitive pressures on other currencies, which may then be forced to devalue to restore their competitiveness. Can successive competitive devaluations explain their sharp depreciation? The answer seems to be no. We calculated the realignment of East Asian currencies versus the dollar necessary to restore the real exchange rate to its June 1995 level under two scenarios: first, the country devalues alone, and the real competitive exchange rate versus the dollar of all other countries remains unchanged; second, all the five East Asian countries affected by the crisis devalue simultaneously. The figure below presents the results of our calculations. If each country devalues alone, the required depreciation vary between 10 and 20 percent. If all five affected countries devalue at the same time, required depreciations increase only by one-half to one percentage points. The difference between the two scenarios is relatively small, and competitive devaluations appear insufficient to explain actual depreciations.

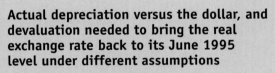

Actual depreciation versus the dollar, and devaluation needed to bring the real exchange rate back to its June 1995 level under different assumptions

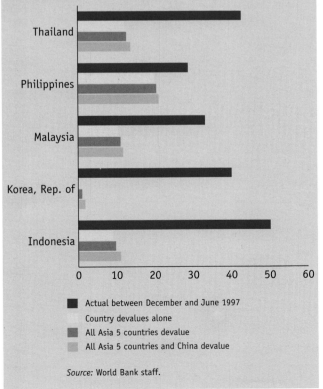

■ Actual between December and June 1997
  Country devalues alone
■ All Asia 5 countries devalue
  All Asia 5 countries and China devalue

Source: World Bank staff.

Stock markets' amplitude and the severity of their boom and bust cycle had, in the past, been less in East Asia than that of emerging markets in Latin America, and had only a slightly more pronounced than those in G7 countries (Kaminsky and Schmukler, 1998). However, the drop in stock prices since July 1997 (in local currency and in dollar terms) has been more severe than in any previous downturn in East Asia, and exceeds downturns in most Latin American countries following the Tequila Crisis. Exchange rate co-movement was very high among the four Southeast Asian countries from August to February, and even higher from October to early January 1998. Since mid-January, there has been less of co-movement in exchange rates as the Malaysia ringgit, the Korean won, and the Thai baht appreciated while the Indonesian rupiah further depreciated. Figures on bond and other capital market instruments for the last quarter of 1997 and the first half of 1998 show that private flows came to a virtual standstill for all East Asian countries.

Trade links between countries meant that declines in demand of imports in one country led to decline in exports in other countries. However, these trade linkages explain only a small portion of the co-movements.[10] Intra-regional exports among East Asian countries accounted for almost 40 percent of total exports in 1996, up from 32 percent in 1990. If Japan is included, the figure rises to 50 percent. These high levels of intra-regional trade reflect a process of specialization and outsourcing of activities from the more advanced, to the lower income countries in the region. About three-fourths of the intra-regional trade is in raw materials, intermediate inputs, and capital equipment which accounts for more than 50 percent of the total East Asian imports of these products. Such trade complementary probably increased the speed and directness of the contagion. As seen in box 1.2, these links were probably more significant mechanisms than "competitive" devaluations.

Other forms of contagion are more difficult to quantify. Financial linkages in the region, including through foreign direct investment, bank lending, and capital market activities, may have meant that events in one country negatively affected another country.[11] Investors from outside the region may have been forced to sell assets in one country in response to losses in other countries. The market structure of foreign exchange

What triggers market jitters?

In the chaotic financial environment of East Asia last year, daily changes—usually falls—in stock prices of as much as 10 percent became commonplace. A recent paper by Kaminsky and Schmukler analyzes the twenty largest one-day swings in stock prices (in U.S. dollars) in Hong Kong (China), Indonesia, Japan, Korea, Malaysia, the Philippines, Singapore, Taiwan (China), and Thailand since January 1997 to see what type of news moves the markets in days of extreme market jitters. Of particular interest was whether news in one country would affect markets in another, and if so, what type of news. Also, did news have a persistent, trend-changing effect or was it merely transitory? The researchers classified news into seven different categories: news related to agreements with international organizations, the financial sector in each country, monetary and fiscal policies, credit-rating agencies, the real sector, and political announcements.

The results? Some of the largest one-day downturns cannot be explained by any apparent substantial news, either economic or political, but seem to be driven by herd instincts of the market itself. Second, most movements are triggered by local news, not news of a country's neighbor or the major Organization for Economic Co-operation and Development (OECD) trading partners. The most weighty events in the market's eyes are news about agreements with international organizations, capital controls, credit rating agencies, financial sector, and political events or rumors. Still, some types of news from abroad have significant effects on the stock market. For example, agreements with international organizations—IMF, the World Bank, and others—have a substantial effect on both home countries and neighbors. In fact, positive news about an agreement in a given country in the region produced, on average, a 9 percent increase in another country's stock market.

How news about local and foreign events affect stock markets

(Percent average daily movement up or down)

Lighter bars indicate insignificant means.

While the analysis focuses on the largest daily movements, the results do not necessarily reflect shocks that are immediately reversed. In fact, the study found that, on average, the one-day market rallies are sustained and trend-changing. This is not the case of market downturns. Within a few days, the largest one-day losses are recovered, suggesting that, on balance, investors overreact to bad news.

Source: Graciela Kaminsky and Sergio Schmukler, 1998, "What Triggers Market Jitters? A Chronicle of the Asian Crisis," World Bank mimeo.

---

trading may have also played a role. Also events in one country may have led market participants to revise their model of development in East Asia more broadly, which may have negatively affected asset prices in a larger group of countries.

Announcements of delays or revisions in multilateral agreements in one country were associated with negative price effects in other countries, particularly in the stock markets. Analysis of movements in exchange rates and stock prices across countries in response to announcements and events in other countries in East Asia reveal that there were important spill overs. These effects were not limited to East Asia only, but also affected some countries outside the region (see box 1.3).

From currency and financial crisis to economic and social crisis

As the Thai crisis spread throughout Southeast Asia, and then to Korea, the currency crisis became a financial crisis, which became, in turn, an economic and social crisis. These have had sweeping domestic political ramifications in Korea and Thailand, where transitions were managed relatively free of disruption. Indonesia experienced an even more profound political crisis.

The outflow of capital in 1997 required huge swings in current account balances. New surpluses were achieved primarily through import compression and reductions in income, rather than through exports (see

"Your money or your exchange rate": Indonesia's monetary dilemma at end-1997

Indonesia's banks have long suffered from well-known weaknesses. Even before the crisis of 1997–98, many banks were insolvent or seriously undercapitalized. But, the central bank lacked the authority to move decisively against the owners, and years of rapid growth had accommodated these serious weaknesses. When the Indonesian rupiah began to depreciate in July 1997, the depositors' confidence was shaken And runs started on many private national banks as early as August 1997.

Later, as part of the first IMF program, the Minister of Finance announced, on November 1, 1997, the closing of 16 small, private banks. Despite the introduction of small-depositor guarantees, depositors panicked and several private banks (including some of the largest) experienced further significant runs. Frightened by the prospect of more bank closures, Bank Indonesia began supplying large amounts of liquidity to keep banks afloat, a move that magnified the macroeconomic crisis.

During the final months of 1997, Bank Indonesia's emergency liquidity support expanded rapidly, and amounted to roughly 150 percent of base money, or 5 times the amount prior to the onset of the crisis. Bank Indonesia sterilized the impact of virtually all of this emergency credit on base money in 1997, resulting in a very stable 12-month growth rate of base money (at around 35 percent) during the second half of 1997. Roughly three-fourths of the emergency credit was provided to the four largest banks, all of which were subsequently taken over by the Indonesia Bank Restructuring Agency. However, extensions of emergency credit reached alarming proportions in early 1998. By the end of February, total liquidity support had doubled compared to the years end, and base money was accelerating significantly (to 70 percent by the end of March 1998 versus less than 35 percent at the end of 1997). Moreover, inflation was rising rapidly (it peaked at a monthly rate of almost 13 percent in February 1998), and the value of the rupiah was sinking like a stone. Indonesia seemed on the brink of slipping into hyper-inflation.

Source: World Bank staff.

figure 1.6). While exports experienced some increase, their performance has not been sufficient to make the current account adjustments without steep economic contraction. Private investment–savings balances have undergone adjustment, and deflated demand.

Monetary and fiscal policy was tightened as countries struggled to cope with the financial panic that had induced a run on their currencies. Rising interest rates were intended to increase the price of domestic assets and make them more attractive to holders of foreign currency funds. But, this raised difficult tradeoffs. High rates compounded the problems of the debt-laden corporate sector. Interest costs rose at exactly the time their profits were falling with the advancing recession, and many could not service their debts. Non-performing loans on the balance sheets of the banks increased, forcing them to call in loans in a struggle to maintain cash flow.

Fiscal policy was set with a view toward off setting the costs of debt restructuring and curbing future inflation. In Thailand, the August 1997 economic program implied a 3 percent of GDP fiscal belt-tightening, with the aim of producing a 1 percent surplus. Indonesia's first program included a 1.2 percent tightening. These targets would not have been strongly contractionary had projected assumptions about economic growth

been realized. For example, growth in Indonesia was programmed to be 5 percent in fiscal year 1997/98 and 3 percent in 1998/99, numbers not inconsistent with private estimates at the time. Because growth did not materialize, the *ex post* fiscal stance was contractionary. Indeed, as the depth of the recession became evident, policy became successively more expansionary (a point to which we return in chapter 7).

Thailand, Korea, Indonesia, and Malaysia have fallen off their growth path into deep recession, and face the prospect of sharp economic contraction in 1998. With the recession in Japan, Hong Kong (China) also seems destined to stagnation in 1998. The Philippines and Singapore have fared somewhat better, but these economies will probably experience "growth recessions"—that is, growth rates so low they will not absorb the increase in employment in the labor markets.

By mid-1998, indicators of real sector activity showed little evidence of recovery. Thailand, the first into recession, may have bottomed out if the floor on the manufacturing and retail sales index holds steady. Korea seems to be bottoming out as well. Malaysia and the Philippines are not yet picking up. Exports have not accelerated, forcing the adjustment onto import compression and output reductions (see figure 1.6). This is

FIGURE 1.7

Trade and production

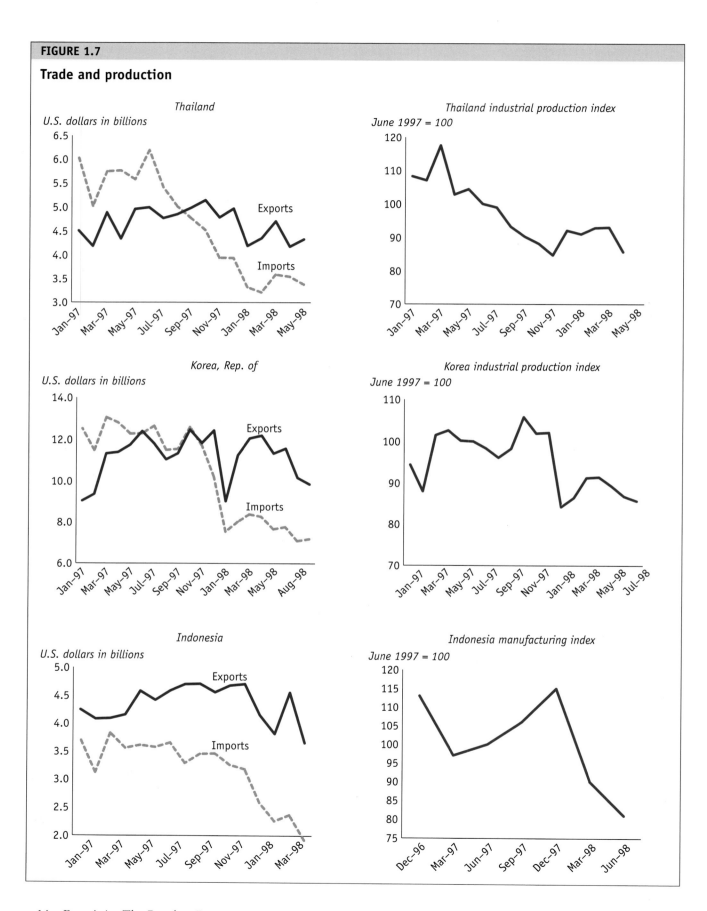

Thailand

U.S. dollars in billions

Thailand industrial production index

June 1997 = 100

Korea, Rep. of

U.S. dollars in billions

Korea industrial production index

June 1997 = 100

Indonesia

U.S. dollars in billions

Indonesia manufacturing index

June 1997 = 100

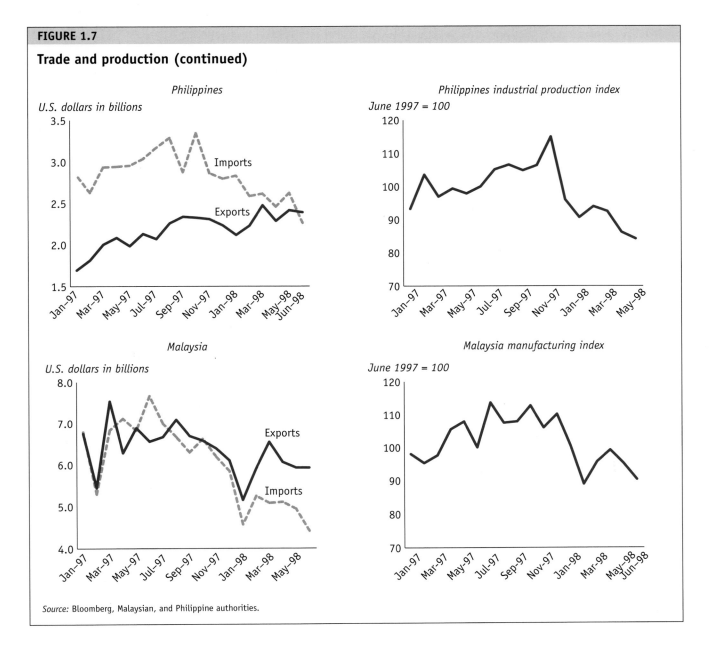

FIGURE 1.7

Trade and production (continued)

Philippines

U.S. dollars in billions

Philippines industrial production index

June 1997 = 100

Malaysia

U.S. dollars in billions

Malaysia manufacturing index

June 1997 = 100

Source: Bloomberg, Malaysian, and Philippine authorities.

in part, because the corporate and financial sectors in the East Asia 5 are hobbled in distress. With interest burdens too heavy to service, the corporates are in no position to invest, even in potentially profitable export industries. The financial sectors, already under cash flow pressures as deposits are withdrawn and loans are going unserviced, have to rebuild their capital to stay in business. Confidence among external investors and domestic investors with holdings abroad has yet to return.

Unemployment has risen as the economies have contracted. In virtually every country in the region, unemployment is at a 20-year high. In some countries, the unemployed have suffered in relative silence; in others, the contraction has led to social protests.

The silver lining of these dark clouds is that financial stability may be taking hold everywhere. Except for Indonesia, exchange rates hovered within a relatively narrow band for the first semester of 1998 and even appreciated since January.

Macroeconomic policy, in the meantime, has to carefully balance the provision of liquidity to a distressed corporate and financial sector with the need to maintain some degree of stability in the exchange rate.

Indonesia responded to this dilemma in late 1997 by providing liquidity to the banks. When combined with the failure to follow through on policy actions, the result was a 50 percent in a drop in the value of the rupiah, which left the country on the verge of hyperinflation (see box 1.4). Most other countries have been able to reduce interest rates, and they have eased downward in Thailand, Malaysia, Korea, and the other major markets of unaffected countries. Nonetheless, the presence of an open capital account and potentially huge pools of capital just offshore undercuts the effectiveness of the domestic monetary policy in any one country.

Conclusions and organization of this study

East Asia's downturn has mixed currency and banking crises with regional financial panic into a particularly virulent crisis. This threatens the very financial viability of its banking and corporate sectors, and dampens prospects for a quick recovery. To be sure, deep-seated structural problems evident prior to 1997—in the financial sectors, the social safety net and environment—required attention if the progress were to be sustained. But countries were working on these problems, (albeit too slowly in retrospect), and the depth of devastation stemming from the economic unraveling that has occurred since July 1997 has gone far beyond what could be attributed to these structural trends.

The inflows of capital, often without due diligence and beyond prudence, were intermediated through poorly regulated domestic financial systems, and fueled a domestic credit boom. This was a series of bets that paid off initially in the form of more rapid growth. But, as inflows pumped up asset prices and increased corporations' leverage, they exposed the region to the shocks of changing investor expectations. In retrospect, rapid growth had spawned structural weaknesses in three dimensions:

- Large current account deficits, financed with short-term flows, exposed East Asian economies to sudden reversals.
- Domestic financial markets' growth and integration with global markets outpaced domestic regulation, and exposed banks to asset–liability mismatches. Inadequate regulation allowed banks and corpo-

rates to assume unhedged foreign borrowing positions and maturity mismatches that left them vulnerable to sudden currency fluctuations.
- Companies, without alternative sources of financing, borrowed heavily from banks to finance their rapid expansion, and in the process became very highly leveraged, which opened them to interest rate surges.

It is important to note that domestic policy failures explain only part of the emergence of these vulnerabilities and subsequent crisis. Failures in international financial markets in the boom part of the cycle, as well as during the crash, were no less pivotal; herd instincts in financial markets have not served East Asia well. Arguably, investors' expectations were made more volatile by the absence of certified data on corporate finance, banks' portfolios, and even the reserve position of the central banks; during the boom years, the absence of data militated against prudential caution, and then during the crash, it fed panic. Similarly, private credit rating agencies and regulators in the OECD, particularly Japan and Europe, failed to raise red flags that might have braked the momentum of overlending. Also, questions have arisen about the early rescue packages led by the multilateral institutions; with benefit of hindsight, were they overly contractional and did they inadvertently push the economies deeper into recession? Were programs too comprehensive and did they overload the capacity of domestic policy-making? Definitive answers to these questions and others await more research. Nonetheless, it is clear that different behavior of private agents and different policies domestically in the one or two years before the crisis might well have pre-empted the crisis or contained it at an earlier stage. The crisis was not pre-ordained.

The more urgent agenda lies ahead. This study focuses on two questions:

What policies can the region adopt to overcome its severe problems and firmly establish recovery? East Asia's crisis is best seen as a story of rapid growth built on an incomplete foundation, which was left exposed to the winds of international capital markets. Now that the financial earthquake has occurred, it will have to rebuild its success on new foundations in its trade competitiveness, in the financial sector, and in the governance and financing of its corporate sectors. The following three chapters go into greater depth, paying particular attention to the links between macroeconomic and microeconomic policies. Each chapter traces the fundamentals of the problem, presents a progress report on actions taken during the first years following the crisis, and offers a view of future policies.

Once in recovery, how can countries ensure its sustainability? Even prior to 1997, deep-seated structural problems—in the corporate and financial sectors, the social safety net and environment—required attention if the progress was to be sustained. Low-income groups and the environment are clearly suffering. While governments have been working to address these problems for years, the depth of devastation stemming from the economic unraveling that has occurred since July 1997 has made their resolution them more pressing. Chapters 5 and 6 focus on the social sectors and environment, and the last chapter speaks to the region's strategy for recovery and economic forces shaping its prospects.

Notes

1. The international poverty line is one dollar a day measured in constant 1985 purchasing power parity terms.

2. Other factors were also important: high government and corporate savings, macroeconomic stability, demographics, and other government policies.

3. Among others, see World Bank, *The East Asia Miracle*, 1993; *Asian Development Bank Emerging Asia*, 1997.

4. Estimates of TFP growth in Asia differ among recent studies, chosen depending on the methodologies, countries, and time periods:

Estimates of TFP Growth in East Asia

	Bosworth & Collins (1996)	Young (1995)	East Asia Miracle (1993)	Sarel (1997)
Hong Kong (China)		2.3	3.4	
Indonesia	0.8		1.5	1.2
Korea	1.5	1.7	2.2	
Malaysia	0.9		0.5	2.0
Philippines	-0.4			-0.8
Singapore	1.5	0.2	2.1	2.2
Taiwan (China)	2.0	2.1	2.7	
Thailand	1.8		1.3	2.0
East Asia	1.1			
United States	0.3			0.3
Other Industrial Countries	1.1			

Source: Bosworth and Collins (1996), Table 7 (covering the period 1960-94), Young (1995), Table 15 (covering the period 1966–90), East Asia Miracle (1993), covering the period 1960-85, and Sarel (1997), covering the period 1978-96.

5. This section benefited from P. Alba, A. Bhattacharya, S. Claessens, S. Ghosh, and L. Hernandez, "Volatility and Contagion in a Financially-Integrated World: Lessons from East Asia's Recent Experience." Paper presented to the PAFTAD 24 conference "Asia Pacific Financial Liberalization and Reform," May 20–22, 1998, Chiangmai, Thailand.

6. Kaminsky and Reinhart, 1997 confirm this finding for a wider set of countries.

7. Macro and currency risk factors constituted 16 percent of the total of the average nominal interest rate of 16 percent over this period, while the base U.S. risk-free rate represented on average 28 percent of the nominal interest costs.

8. With the possible exception of Thailand, foreign investors apparently did not play a large role in triggering the crises. Data on mutual and pension funds' holdings of equities and other assets in East Asian countries do not suggest a massive outflow of foreign capital during the July–December 1997 period. Foreign investors appear to have reduced their holdings prior to the crisis, particularly in case of Thailand, and increased their holdings, particularly in the first few months of 1998. Hedge funds and other short-term investors played a limited role in triggering the crisis (see Eichengreen and Mathieson). The massive reversal in capital flows arose mainly from (a) the reluctance of foreign lenders to roll over short-term claims after September 1997 (causing an outflow of about US$50 billion for the affected countries), and (b) the purchases of foreign exchange by local corporations to cover open positions, and in Indonesia capital flight.

9. It is important to distinguish pure spill overs from co-movements that are due to similarities in changes in underlying fundamentals. In practice, however, this is difficult. Asset prices, for example, reflect market expectations of future real returns, which depend on expectations about fundamental economic variables and the market perception of risks and its willingness to absorb them. If expected fundamentals and risks changed in similar ways for all East Asian countries during this period, co-movements may arise as rational, market responses. Similarly, capital flows may exhibit co-movements due to similar changes in fundamentals.

10. McKibbon and Martin, 1998, conclude from their modeling that "...contagion did not arise through direct trade or capital account linkages" (1998:43). See also Hoekman and Martin, 1998.

11. See Calvo and Reinhart for evidence on Latin America following the 1995 Mexico crisis.

chapter two

Trade and Competition

Two freighters carrying logs from the Weyerhaeuser Corporation in the Pacific Northwest steam idly at sea near South Korea, their cargoes undelivered—perhaps undeliverable. Just as the ships approached port, the Korean buyers refused to pay for wood they no longer needed to build homes and furniture they can no longer sell while Asia's financial crisis spreads.

Tiffany & Co., enjoying brisk jewelry sales on Fifth Avenue, is hurting in Hawaii; the Japanese tourists who normally visit and buy there are doing neither lately. And Reebok International, the shoe giant, reports that its sales to Asia's consumers are way off. Take the Rockport shoe boutique, a Reebok subsidiary, in Seoul's upscale Myondong neighborhood. With few customers to serve, sales clerks spend their time shining the shoes on display.—Louis Uchitelle, "Dim Asian Economies Casting Shadows in the U.S.," *New York Times*, December 14, 1997.

Trade has been the engine of growth for East Asia's 30-year rapid expansion. To promote export development, governments in East Asia initially provided various incentives, such as duty rebates, input, and credit facilities with preferential lending rates. These policy interventions were motivated by the belief that shifting industrial structures toward newer and more modern sectors increases the opportunities for capturing dynamic scale economies. Later, reliance on more generalized incentives grew, which included reforms of trade and investment regimes, appropriate exchange rates, and macroeconomic policies. Correspondingly, trade as a share of gross domestic product (GDP) increased considerably, from 15 percent in 1970 to over 50 percent in 1995. Exports grew by over 10 percent per annum in virtually every quarter. Between 1970 and 1995, per capita exports increased from US$100 to US$400 in the Republic of Korea and from US$80 to US$850 in Thailand.

Then, in 1996, trade crashed. The magnitude of this deceleration was unprecedented in recent history. From the first quarter of 1995, when export growth among the East Asia 5 (Thailand, Korea, Indonesia, Malaysia, and the Philippines) and other East Asian countries reached a peak, growth started to decelerate dramatically (see figure 2.1). By the first quarter of 1996, export growth fell to zero in the East Asia 5 countries and to negative rates for other East Asian countries, including China and the Newly Industrialized Economies (NIEs). As noted in chapter 1, these events coincided with rising external and internal vulnerabilities in these countries. As the news spread of slowing export growth in all East Asian countries—a region where trade was a more important engine of growth than elsewhere in the world—the fears about external conditions and competitiveness issues became more significant.

This chapter asks the questions: Is the downturn in exports a cyclical problem that can be resolved rela-

tively quickly through exchange rate adjustments or is it a deeper structural problem? Can exports lead the recovery if the region remains mired in recession?

Causes of the 1996 export slowdown

The recent slowdown in export growth reflects forces that largely are cyclical in the world and within the regional economy. These include:

- A large fall in world trade growth
- Yen depreciation in Japan
- Real effective exchange rate appreciations in some East Asian countries and
- Significant price declines for major export products in some countries in the region.

The fall in world export growth from its cyclical peak in 1995 was the largest in the past 15 years—from about 20 percent to about 4 percent in U.S. dollars in just one year. Within East Asia, export growth fell by the same proportion, although the extent of the slowdown varied across countries (see table 2.1). Within the East Asia 5, Thailand was the worst affected, recording negative export growth in nominal terms in 1996, followed by Korea. Since 1996, export growth has remained slow throughout the region, with the exception of the Philippines and China.

FIGURE 2.1

World, East Asia, and countries' growth 1982–97

(dollar, percent)

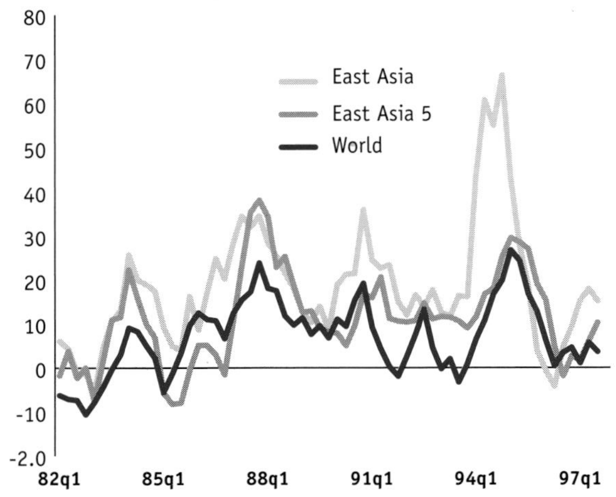

Source: IMF, IFS.

TABLE 2.1

Recent export slowdown in East Asia
(dollar percent)

	1994	1995	1996	1997
Thailand	19	20	-1	3
Korea, Rep. of	14	23	4	5
Indonesia	8	12	9	7
Malaysia	20	21	6	1
Philippines	17	24	14	21
China	25	19	2	21
Hong Kong (China)	11	13	4	4
Singapore	24	18	5	-1
Taiwan (China)	9	17	4	4
EA9	19	21	4	8
Japan	9	10	-8	2
United States	9	12	7	10
World	14	20	4	4

Source: IMF, IFS, staff estimates.

The sharp depreciation of the yen in 1995 compounded the negative impact of the slowdown in world exports on many East Asian countries. Japan is both a major market for other East Asian producers and a competitor in export markets. Although the previous appreciation of the yen sent import growth soaring between 1992 and 1994, the subsequent reverse sharp depreciation sent the real value of imports plummeting. This has mainly affected regional countries whose export structure is similar to Japan's, such as Korea and Hong Kong, China. In 1996, Japan's imports from

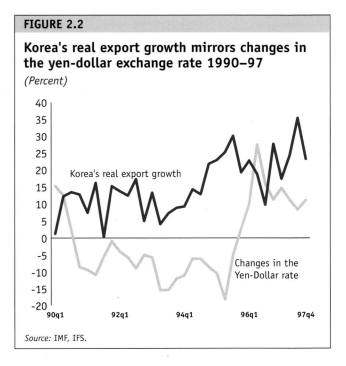

FIGURE 2.2

Korea's real export growth mirrors changes in the yen-dollar exchange rate 1990–97

(Percent)

Source: IMF, IFS.

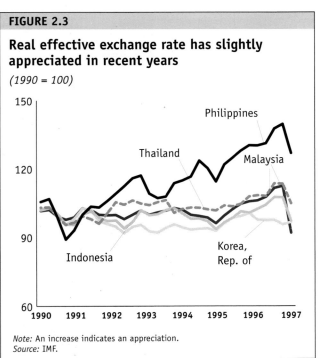

FIGURE 2.3

Real effective exchange rate has slightly appreciated in recent years

(1990 = 100)

Note: An increase indicates an appreciation.
Source: IMF.

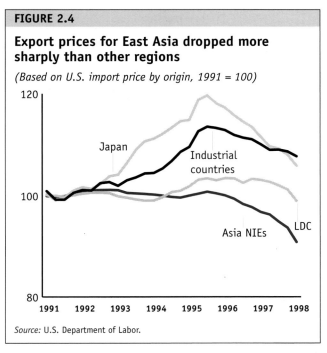

FIGURE 2.4

Export prices for East Asia dropped more sharply than other regions

(Based on U.S. import price by origin, 1991 = 100)

Source: U.S. Department of Labor.

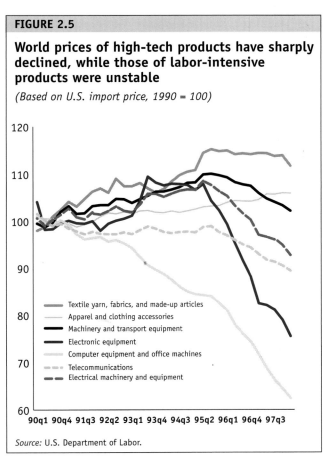

FIGURE 2.5

World prices of high-tech products have sharply declined, while those of labor-intensive products were unstable

(Based on U.S. import price, 1990 = 100)

- Textile yarn, fabrics, and made-up articles
- Apparel and clothing accessories
- Machinery and transport equipment
- Electronic equipment
- Computer equipment and office machines
- Telecommunications
- Electrical machinery and equipment

Source: U.S. Department of Labor.

Korea fell by 8.5 percent and imports from Hong Kong, China fell by 6.6 percent, while its imports from labor-intensive China rose by 10 percent and from the Philippines by more than 23 percent. The depreciation of the yen also significantly affected regional industrial exports to the markets of underdeveloped economies, where the more industrialized countries of East Asia—Hong Kong (China), Taiwan (China), Singapore, and Korea—compete against Japan. This was especially true with Korea, whose real export growth has tended to mirror changes in the yen–dollar exchange rate, rising with an appreciation of the yen and falling with its depreciation (see figure 2.2).

Although effective exchange rates were stable in the region during the 1990s, certain East Asian countries started to experience some appreciation of the real effective exchange rate beginning in mid-1995 until the second quarter of 1997 (see figure 2.3), including the Philippines (15 percent), Thailand (12 percent), and Indonesia (11 percent). This appreciation may have hurt exports in those particular countries.

Some Asian economies were also hit with significant price declines of their major export products. Based on U.S. import prices by origin, the sharpest fall in prices was for imports from Asian countries, which dropped by 25 percent relative to industrial country import prices (see figure 2.4). Within East Asia, only the Philippines had rising export prices at the onset of the crisis.

The largest price declines occurred in the electronics industry, especially for computers, semiconductors, and telecommunications, in which East Asia specializes. Korea was particularly hard hit when the 16MB DRAM chips, which account for a large share of its electronics exports, fell from a peak of about US$150 per unit in 1993 to less than US$10 in 1997. Prices of labor-intensive manufactured goods, such as textiles and apparel, have been more stable—one important reason that China's export growth, which is still heavily based on labor-intensive manufacturing, remains strong (see figure 2.5).

Cyclical or structural?

Although the sharp decline in East Asian export growth in 1996 and later can be attributed largely to the unusual and unfortunate confluence of several adverse cyclical factors, there may have been additional structural or competitiveness problems underlying the slowdown. When the composition and evolution of the export system in East Asia is studied, some fragilities are revealed.

One of the most remarkable aspects of East Asia's export performance has been the rapid shift in the composition of exports from resource- and labor-intensive industries to more skill- and capital-intensive industries (see figure 2.6). In 1990, most East Asian countries were still heavily dependent on resource-based exports, which accounted for about 40 percent of total exports in China, Thailand, and Singapore; 54 percent in Malaysia; and as much as 72 percent in Indonesia. Hong Kong (China), Korea, and Taiwan (China) depended heavily on low-technology products—about 60 percent of their exports. At the same time, about 30 percent of Malaysia's and Singapore's exports were in high-technology products, whereas China did not

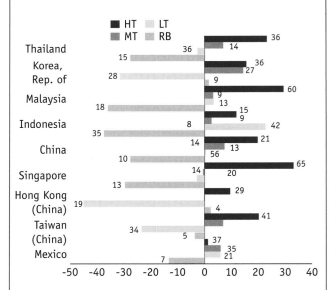

FIGURE 2.6

The dramatic change in export composition in East Asia

(change in shares of exports by category in total exports between 1990 and 1996)

Note: Resource-based (RB) products refer to processed foods and tobacco, simple wood products, refined petroleum products, dyes, leather (not leather products), rubber products, and organic chemicals; low technology (LT) products refer to textiles, garments, footwear and other leather products, toys, simple metal and plastic products, furniture, and glassware; medium technology (MT) products refer to automotive products, chemicals, industrial machinery, and simple electrical products; high technology (HT) products refer to fine chemicals, pharmaceuticals, complex electrical and electronic machinery, and precision instruments.

The numbers near the bars refer to the actual share in total exports of the corresponding category in 1996.

export such products. By 1996, there was a dramatic change in the export structure in East Asia. Singapore and Malaysia doubled their share of exports in high-technology products, while Thailand almost tripled its share of exports in high technology. Indonesia reduced its share of resource-based exports but is still low on the technology ladder for most exports. In comparison, Mexico's export structure changed little during the same time period.

This situation is not a picture of a collapse in competitiveness or other structural problems; it is a picture of rapid trade- and investment-led gains in market share and convergence, exactly what would theoretically occur. What is unusual about it is the speed of the transformation of the export structure.

What is behind this phenomenon? Although rapid structural transformation in East Asia has been associated with a much faster accumulation of human and physical capital per capita than other developing countries, in recent years both "push" and "pull" factors have operated to hasten such structural transformation in East Asian countries compared with other developing regions. First, competition in the labor-intensive goods market from low-cost producers such as China may have been an important push factor, especially in recent years. Second, foreign direct investment from industrialized countries was the driving force behind the pull factor in the second generation of East Asian NIEs. Third, the relocation of the production bases of the first generation of NIEs (Korea, Hong Kong, China, Taiwan, China, and Singapore) to neighboring countries in the late 1980s as their wages and costs rose also reinforced the pull factor.

As a result of these factors, export structure in East Asia has become characterized by (a) early exit from low-skill labor-intensive exports; (b) specialization in high-technology exports, mainly electronics; and (c) strong intraregional links. These characteristics have caused East Asian economies to become increasingly dependent on one another and on fewer products as their exports. Although these countries have demonstrated the ability to adapt to external changes in past decades, this new export structure may have exposed them to greater risk of export instability and "contagion."

Competition from China and other low-cost exporters

China's rapid entry into world markets may have accelerated the early exit from labor-intensive exports of lower-income Asian countries. Structural reforms that China has undertaken in recent years, together with the self-reinforcing effects of foreign direct investment and exports, have significantly improved China's international competitiveness and export performance. Exchange rate policies, however, did not cause any notable difference in competitiveness (for China, see box 2.1).

Figure 2.7 shows China's share of the world market for those product groups (measured as two-digit Standard International Trade Classification [SITC] categories) that were the top 10 manufactured exports of Indonesia, Malaysia, and Thailand, respectively, from 1988–90. Malaysia and Thailand continued to increase their world market shares of their top 10 exports but at a much slower pace than that of China. Indonesia lost world market share in those products.

The decline in the growth of market share for Indonesia, Malaysia, and Thailand's top products may be due to several factors. It could reflect a normal transition to higher-end products, which would have happened independently of China's actions, or it might truly represent a market displacement by lower-cost competitors. While it is difficult to distinguish between these two effects, statistical analysis shows that during 1989–92, increases in China's world market shares were not significantly associated with declines in world market shares for these products in Indonesia, Malaysia, and Thailand; however, recent growth of China's market share in its top 10 manufactured exports was associated with a small decline in the market share of these products for Malaysia and Thailand (see table 2.2).[1]

TABLE 2.2
East Asia makes room for China
(elasticity of East Asia's market share to China's)

	1989–92		1993–96	
Indonesia	0.167	(3.98)	-0.070	(-1.01)
Malaysia	-0.041	(-1.24)	-0.087	(-2.99)*
Thailand	-0.091	(-1.11)	-0.128	(-2.57)*

Note: t-statistics in parentheses.
* Denotes significance at 5 percent level.
Source: Bhattacharya, Ghosh, and Jensen, 1998.

Thailand, in particular, is losing competitiveness in low-cost exports. More than 35 percent of Thailand's exports still consist of low-technology products such as

FIGURE 2.7

Competition from China

Export performance of Indonesia and China in products which were Indonesia's top 10 exports of manufactures in 1988–90
(share of world exports of these products)

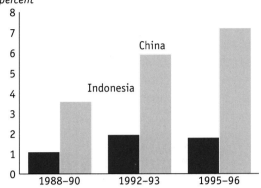

Export performance of Malaysia and China in products which were Indonesia's top 10 exports of manufactures in 1988–90
(share of world exports of these products)

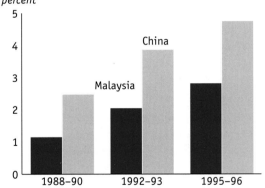

Export performance of Thailand and China in products which were Indonesia's top 10 exports of manufactures in 1988–90
(share of world exports of these products)

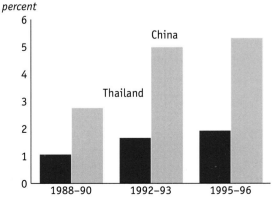

Source: Bhattacharya, Ghosh, and Jansen (1998).

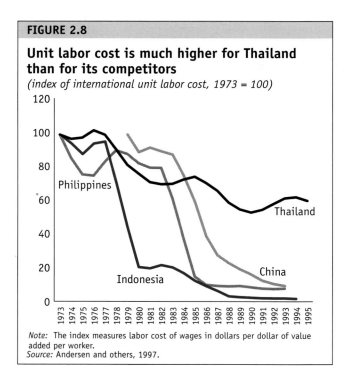

FIGURE 2.8

Unit labor cost is much higher for Thailand than for its competitors

(index of international unit labor cost, 1973 = 100)

Note: The index measures labor cost of wages in dollars per dollar of value added per worker.
Source: Andersen and others, 1997.

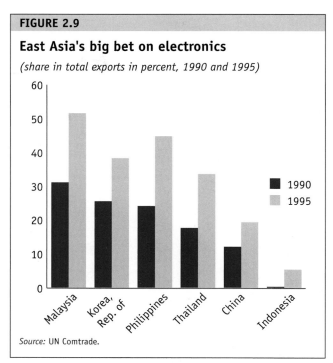

FIGURE 2.9

East Asia's big bet on electronics

(share in total exports in percent, 1990 and 1995)

Source: UN Comtrade.

textiles, garments, and toys. This exposes a large share of its exports to growing low-wage competition from countries such as China, where low-technology products comprise more than 50 percent of its exports. The unit labor cost in Thailand has been much higher than its close competitors without a corresponding increase in productivity (see figure 2.8). In apparel manufacturing, the hourly labor cost was US$0.25 in China in 1995, whereas it was US$1.11 in Thailand.

Narrow specialization in the electronics industry

One of the more striking aspects of East Asia's changing export structure and its increased specialization between 1990 and 1995 has been the dramatic growth in the electronics industry in all East Asian countries, and especially among the East Asia 5. All of these countries showed a significant jump in the share of total exports, including electronics: more than 50 percent of total exports in Malaysia, over 45 percent in the Philippines, close to 40 percent in Korea, and nearly one-third in Thailand (see figure 2.9). China and Indonesia showed far smaller shares of total exports, but the rapid rise in their shares were also impressive. Globalization of production—reducing of the value chain to discrete and specialized steps—has been a key

factor behind East Asia's rapid rise in electronic exports.[2] Thus, 64 percent of the world's production of hard disk drives (HDD) now takes place in Southeast Asia, with concentration in Singapore, Malaysia, and Thailand, and 38 percent of global production of semiconductors takes place in East Asia.

Price wars and intense competition are two features of the global electronics industry that arise from (a) standardization and mass production, leaving less room for product diversification; and (b) a persistent trend toward over-capacity problems. East Asian countries have been aggressive in expanding capacity and capturing significant market shares, but in the process, they also became more vulnerable to terms of trade declines and over-capacity problems.[3] The Korean strategy (see box 2.2) is an extreme example of risk taking by national firms as part of an industrial development strategy. Thailand represents a more intermediate case than Korea, as easy money and financing from abroad encouraged local firms to bet big in the electronics industry as independent second-tier suppliers to multinational corporations. However, it exposed them to over-capacity problems, intense price wars, and vulnerability to large risks.

Korea and Thailand developed national firms in the electronics industry, but Malaysia and the Philippines had a safer strategy: to be part of a more diverse international production network through direct connections with multinational firms. Multinational companies have global brands, their own research and development (R&D) capabilities, and access to global capital, all of which cushion them from market risks. But Korea and Thailand's national firms were exposed to problems of massive investment requirements for entry, over capacity, and price risks. Some indirect evidence for these problems is shown in unit price trends for overall exports (see figure 2.10). Since 1995, Korea has experienced the largest declines in the unit price for its exports, followed by Thailand. The Philippines, in contrast, actually showed a sharp improvement in the unit prices of its exports before a recent downturn. The Philippines has had rising export earnings because it has relied heavily on foreign investment of multinationals, whereas Korea and Thailand chose not to do so and consequently, now face serious economic problems.

Intra-Asian trade: The domino effect

For decades, trade has been the engine of growth in East Asia and, more recently, intra-regional trade has made an increasing contribution to growth. In 1996, the share of intra-regional trade accounted for about 40 percent of total exports, up from 32 percent in 1990

FIGURE 2.10

Comparison of Philippine, Thai, and Korean export prices

(1990 = 100)

Source: IMF, IFS, Datastream.

BOX 2.2

Korea: Big bets, big losses

Korea's success in the global electronics industry has been largely due to big bets by national firms and is closely tied to the semiconductor industry development, the role of chaebols, and industrial policy. In the 1990s, Korea's semiconductor industry grew dramatically, expanding the country's share from virtually nothing to one-third of the global DRAM market. However, its rapid expansion also created several vulnerabilities: (a) 80 percent of Korea's total semiconductor production was narrowly focused on DRAM chips, (b) the industry relied heavily on imported manufacturing equipment and intermediates, and (c) it relied on massive investments through major chaebols' access to easy credit from the domestic financial system.

By 1996, the 4MB and 16MB DRAM chips suffered a price collapse brought on by Korean firms' over-investment and over-capacity, in addition to technological advances in higher memory chips. Prior to the collapse, Korea heavily increased capacity, ignoring the softening demand, emerging supply surplus, and arrival of next-generation chips. A distinguishing feature of Korea's strategy was to combine high investment thresholds with extremely volatile income streams, leading to boom–bust cycles. The chaebols could embark such a strategy because they had access to cheap capital and presumed ability to bear risks inter-

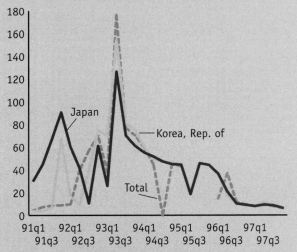

U.S. import deflator of 16 MB DRAM chips

Source: U.S. Bureau of the Census, 1998.

nally, but in the end, big bets in a narrow product range of DRAM chips led to major losses.

Source: Kishimoto, Nomura Research Institute, 1998.

TABLE 2.3
East Asia's intra-regional exports by country, 1996

(Share of total exports in percent)

	China	Hong Kong (China)	Indonesia	Korea, Rep. of	Malaysia	Philippines	Singapore	Thailand	Taiwan (China)	Japan	East Asia
China	0	24	1	4	1	1	2	1	2	19	56
Hong Kong (China)	27	0	1	1	1	1	5	1	3	5	47
Indonesia	4	4	0	6	2	1	8	2	4	27	58
Korea, Rep. of	7	8	2	0	2	1	5	2	3	14	45
Malaysia	3	5	1	3	0	1	20	4	3	13	53
Philippines	1	5	1	2	2	0	5	5	3	16	40
Singapore	2	9	1	3	19	2	0	6	4	8	53
Thailand	3	5	1	1	3	1	14	0	2	17	48
Taiwan (China)	13	23	2	2	3	1	4	3	0	12	63
Japan	5	6	2	7	4	2	5	4	7	0	42
East Asia	5	10	2	4	4	1	6	3	4	9	49

Source: U.N. Comtrade.

TABLE 2.4
Export share correlations in intra-regional and extra-regional trade, 1996 (above the diagonal: correlations in world market; below the diagonal: correlations in regional market).

	China	Hong Kong (China)	Indonesia	Korea, Rep. of	Malaysia	Philippines	Singapore	Taiwan (China)	Japan
China	1.00	.85	.53	.35	.40	.54	.26	.41	.29
Hong Kong (China)	.81	1.00	.72	.23	.31	.58	.16	.25	.63
Indonesia	.20	.34	1.00	.28	.34	.31	.10	.16	.17
Korea, Rep. of	.35	.69	.28	1.00	.67	.72	.50	.55	.80
Malaysia	.36	.69	.37	.78	1.00	.82	.76	.74	.77
Philippines	.32	.64	.14	.76	.92	1.00	.64	.65	.73
Singapore	.36	.66	.32	.79	.91	.88	1.00	.94	.79
Taiwan (China)	.43	.70	.32	.90	.77	.72	.76	1.00	.81
Japan	.15	.06	.01	.78	.43	.38	.44	.48	1.00

Source: U.N. Comtrade (three-digit SITC, 174 product categories).

(see table 2.3). If Japan is included, the share of intra-regional trade rises to 50 percent. This concentration of trade within East Asia reflects an ongoing process of specialization among countries in the region. The country with the lowest dependence on trade with other East Asian countries is the Philippines, which sends only 25 percent of its total exports to the other Asia 9, which explains the Philippines' stronger export performance after the crisis. Singapore, Hong Kong (China), and Malaysia have the greatest dependence, 40 to 45 percent, on trade with other East Asian countries. Japan is a major market for most of the East Asian economies, with the exception of Hong Kong (China) and Singapore.

The magnitude and interdependency of these trade links was one of the features of the "Asian Miracle" that fueled rapid regional growth. But after the 1997 crisis, these links became a liability because they provided a perfect channel for the contagion to spread swiftly throughout East Asia. It is essential that intra-regional trade links once again become both an asset and a means to spread economic recovery.

To what extent do East Asian exports compete with each other and to what extent do they complement each other? Export share correlations reveal a mixed picture (see table 2.4). Low correlations reflect different endowments of natural resources and patterns of specialization, whereas high correlations reflect similar export structures. Indonesia and China have low correlations with the rest of the region. Japan exhibits a high correlation with Korea because the two countries compete with each other in the regional market and in the world market. On the other hand, exports of China and the Philippines have very low correlations with those of Japan and thus, are complementary. This explains why China and the Philippines have continued to export to Japan after the depreciation of the yen, whereas Korea's exports to Japan have decelerated.

High correlations among some countries, such as Hong Kong (China)–China or Singapore–Malaysia–Philippines, are more likely to reflect intra-industry trade than competition. About three-quarters of intra-regional trade is in intermediates and capital goods,[4] suggesting that a significant part of the trade is complementary. This is a reflection of the first generation of NIEs relocating to neighboring countries, which should make intra-regional trade more resilient to domestic demand shocks than if it were oriented to final goods. But by the same token, these characteristics have amplified the susceptibility of intra-Asian trade to world demand shocks. Because significant parts of each country's trade also competes with one another in world markets, a downturn in world export growth (combined with recent large depreciations) makes it more likely that large terms of trade losses can occur.

To the extent that intra-regional trade was driven largely by a common export platform-based process linked to relocation of labor-intensive industries, and not by a deeper process of true economic integration, the gains from intra-regional trade were less than their potential. A particular aspect of this weakness in intra-regional trade is that all East Asian countries have maintained selective and highly protected domestic industries, such as petrochemicals, steel, and automobiles, that have discouraged deeper economic integration. These factors may explain the weakness of intra-Asian trade flows, especially in 1996 and more recently.

Asian exports in the aftermath of the crisis

East Asian exports, especially in the crisis countries, have turned out to be sluggish even with the stimulus of real depreciation—about 40 percent in Thailand, 57 percent in Korea, 46 percent in the Philippines, and 55 percent in Malaysia between July 1997 and July 1998 (the exchange rate continues to fluctuate in Indonesia). Part of the explanation for this difficult recovery lies in the structure of intra-regional trade. Because East Asian countries compete with and complement one another, the devaluations did not have the expected effect.

Export volume response. A 40 percent depreciation should have increased export volume in the crisis countries by 20 to 30 percent based on typical elasticities.[5] Until April 1998, only Korean export volumes have matched this expectation, expanding by about 30 percent (annualized rate). Export volume growth in both the Philippines and China has held up at above 20 percent (annualized rate). However, the overall response of Thailand's export volumes has been disappointing, despite an initial spurt. In the first half of 1998, they were estimated at only 7 to 8 percent; Malaysia's exports were estimated at similar levels. By way of comparison, Mexico's export volume growth was in excess of 30 percent following the peso devaluation of a similar magnitude.

The mixed picture of the expected versus actual export responses can largely be explained by the complex intra-regional links. As seen in table 2.4, Thailand and Malaysia depend heavily on Singapore as an export market, Korea is competing with Japan, and the Philippines is only moderately dependent on the region. Because approximately 50 percent of export is intra-regional, the temporary collapse in growth within East Asia caused a significant loss in export volumes. Import demand fell by 25 percent in crisis countries, became negative in Japan, and slowed sharply elsewhere, even in China. As a result, intra-regional export volumes have actually shrunk by about 5 percent.

Other factors have influenced the sluggish response, some resulting directly from the crisis and others already existing but only now becoming apparent. The credit crunch in the crisis countries and the loss of trade credits associated with the financial crisis would also be expected to affect the supply response (and import demand) in the crisis countries. In a survey of exporting firms in Thailand, other constraints have been unveiled: physical infrastructure limitations and a skilled labor shortage (see box 2.3).

Export prices. The volume gains in Korea have been almost fully offset by price declines in the unit value of

Constraints to exports in Thailand

The challenges facing Thai firms were highlighted in a study conducted between November 1997 and April 1998. Over 1,200 firms were interviewed to learn about the impact of the crisis on their performance, the difficulties they face in restructuring, and the long term problems in improving their competitive position. Almost 70 percent of Thai firms are operating at a lower capacity than they were in January 1997 and more than 50 percent have reduced their workforces. Exporting firms have faired somewhat better, but still over 40 percent of them now employ fewer workers. About 30 percent of exporters believe that they will be increasing their capacity within a year.

Most Thai firms report that demand decline is the primary difficulty they are facing; a close second is the depreciation of the baht raising the costs of inputs. As a result, firms are facing a cash flow squeeze; finance is a problem and higher interest rates make it difficult to service loans. Given the fall in demand and higher interest rates, most firms are not seeking additional credit, but about one-third of firms do claim that they have inadequate access to finances. According to the survey, the largest challenge is to sell the goods, rather than to find the money to produce more goods.

In the long term, exporting firms report more difficulty in increasing their productivity than nonexporting firms. Not surprisingly, they perceive customs administration, corruption, and bureaucratic red tape as serious bottlenecks. Businesses see finance as a moderately severe constraint, and one of the biggest challenges is getting access to long-term credit. Currently, over 80 percent of loans are due within a year. This not only creates a mismatch between the terms of credit and the projects under-

taken by firms, but also limits firms' abilities to shelter themselves from volatility in the cost of capital. Although the decline of the baht was predicted to make Thailand more competitive, most Thai firms surveyed see the labor costs as a severe hindrance to increasing production.

Source: Dollar and Hallward-Driemeier, 1998.

Bottlenecks facing Thai firms
(firm rating on a scale of 1=no problem to 5=severe problem)

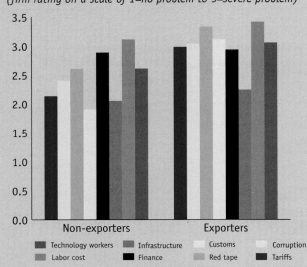

Note: The numbers are based on a rating by survey firms about their perception of bottlenecks to production and productivity growth, rating them on a scale of 1=no problem to 5=severe problem.
Source: Dollar and Hallward-Driemeier, 1998.

exports by about 20 percent of the annualized rate. Overall, U.S. import data show that the pace of decline in prices from Asian NIEs is accelerating—from 6 percent annually in the third quarter of 1997 to 12 percent annually in the first quarter of 1998. A significant part of this decline is due to the rise in the dollar's value, which is dragging down all export prices in world trade, not just Asian export prices.[6] As counterpart to this, the fall in the value of the yen against the dollar affects prices in the Asian countries and, most directly, Korea.

The fall in Asian export prices also could be due to competition in relative prices. Asian countries tend to compete more intensely with one another than with other world exporters, partly because of greater similarity of export structures.[7] This implies that if all East Asian countries lowered their export prices simultaneously, no one country would increase market share or

export growth and the main effect would be lower export prices. This may be a significant factor in the crisis countries, especially in the electronics sector. A similar effect is observed in agriculture exports (see box 2.5).

In contrast, following the 1995 peso crisis, unit export prices did not decline significantly during Mexico's phase of rapid export volume growth. There are three reasons for this occurrence: (a) the dollar value of world trade was expanding rapidly in 1995, unlike now; (b) much of Mexican rapid export growth was intrafirm exports to the United States from maquiladora factories; and (c) there was no severe generalized regional crisis, as in East Asia.

Singapore's crucial role in the East Asia crisis

Although Singapore has weathered the regional currency crisis reasonably well, the sharp fall in economic activity in the rest of the region has begun to take its toll. GDP growth is expected to slow significantly in 1998 from a high of 7.8 percent in 1997. In the first quarter, real GDP fell by 1.4 percent compared with the last quarter of 1997; this reflects a steep decline in manufacturing and commerce. In late June 1998, the government revised its official forecast substantially downwards, from -0.5 percent to -1.5 percent. To stimulate the economy, the government has announced property tax rebates and an increase in infrastructure spending.

Singapore's future is closely bound with that of its neighbors'. It is a highly open economy; its trade is more than four times its GDP, and its nonresident consumption is 20 percent of its total private consumption. Singapore receives about 20 percent of Malaysia's exports, 14 percent of Thailand's, and 8 percent of

Indonesia's. Some exports are consumed locally, but a large chunk is being re-exported—about half of Singapore's total exports. Singapore is among the largest investors in Malaysia, Thailand, and Indonesia and, in 1996, its share of foreign direct investment in those economies was estimated to be 14 percent. In Malaysia, Singapore's investment is equal to that of Japan's. As a result, since the crisis broke in Thailand and spilled over to Malaysia and Indonesia, Singapore has been working closely with the crisis-stricken countries. It proposed to create a multilateral forum to guarantee Indonesian banks' letters of credit. In April 1998, Singapore detailed a plan to extend a US$5 billion trade finance guarantee to Indonesia. To continue playing the role of a regional hub providing trade, shipping, and port services, Singapore may find it beneficial to expand its coordinating role.

Source: World Bank.

Prospects and policies

Near-term prospects and issues. Given the sluggish growth in nominal dollar world trade, the continued strength of the U.S. dollar, the slowing of both growth and imports in Japan, and the collapse of intra-Asian trade, exports are unlikely to recover quickly. Indeed, there may be further adverse shocks that could destabilize the region's economy. This risk calls for urgent attention at the highest policy levels within the region.

Asian exports depend directly or indirectly on world trade growth. Any additional shocks to the value of the dollar against other currencies would cause increasing instability. Any strengthening would tend to drive dollar prices further down. The yen–dollar exchange rate is also of special interest, and any shocks to it would adversely affect recovery prospects.

In addition to exchange rates, measures are required to revitalize real world trade and import demand in industrialized countries. It is vital to follow the policy of global opening of markets rather than regionalism, especially in Western Europe, where the emerging recovery would strengthen world demand, but only if it were open to all markets in the world. The United States is already helping to maintain open markets and sustain a larger volume of imports, which is crucial to recovery in Asia. Japan has announced a fiscal stimulus, but it must ensure that the stimulus is effective in order

to boost growth and imports. The speed and success of recovery in East Asia critically depends on the ability of Europe, the United States, and Japan to accommodate large trade imbalances.

Trade credits may also help to establish a recovery and would likely increase intra-Asian demand. Japan and other high-income countries in the region, such as Singapore (see box 2.4), could help to implement these initiatives. Encouraging foreign direct investment inflows to the crisis countries, especially in the tradables sectors, would be a vital domestic policy instrument to spur exports and trade. Resolving the financial crisis and banking system problems is essential, especially to ensure access to loans by credit-worthy firms in exporting sectors.

In the medium term. Even when cyclical conditions improve in world trade and the exports from East Asian countries recover, underlying core policy issues will still require attention. First, there is the need to avoid implicit industrial policies and to improve not only financial sector discipline but also corporate accounting and governance rules. Second, domestic policies should try to attract foreign direct investment back into export sectors, not only for immediate capital investment, but also to reduce risks and to increase success in operating in the new global production climate. Third, to encourage deeper and less vulnerable intra-Asian trade, trade barriers among countries need to be lowered sharply,

BOX 2.5

Thailand's agriculture: Producing more, eating less

Prior to the East Asian crisis in 1997, Thailand's terms of trade were generally robust. Since the outbreak of the crisis, however, it appears from the limited available data that the overall terms of trade have deteriorated sharply. While export volume was estimated to have grown by 16.6 percent between March 1997 and March 1998, export value declined by 3.5 percent because of a decline in unit prices of 16.6 percent (Bank of Thailand, 1998). This deterioration in the terms of trade makes it more difficult for Thailand to achieve the turnaround in its current account needed as part of the adjustment to the crisis.

Much of the price decline for manufactured items that are differentiated products presumably results from the need for exporters to accept lower prices if they are to increase their own share of world markets. Bulk agricultural products such as rice, rubber, and maize are typically viewed as homogeneous goods that trade at more or less the same prices worldwide, except for differences resulting from transport costs and other factors. However, the East Asian crisis has been such a shock to the entire world economy that it has had a large impact on world prices for a number of commodities, such as rubber, where the crisis countries are important producers. Furthermore, even commodities that appear to be homogeneous frequently include a variety of characteristics that allow for substantial differences in the prices of the goods in world markets.

The quantity adjustment in Thai agricultural exports is also an important aspect of the country's terms of trade. Since the crisis began in the second half of 1997, there has been little opportunity for producers to adjust their output decisions in response to the new price incentives. Substantial increases in inputs, such as fertilizer and water, occurred during the second rice crop in late 1997. Although Thailand was not as severely affected by the El Niño-induced drought as Indonesia was, dry weather conditions did depress the yields of a number of permanent crops, such as longan and durian. Of course, even if the quantity produced does not change, the quantity exported may increase substantially as a consequence of reductions in domestic demand following the sharp rises in consumer prices.

A simple initial indication of the nature of the price and quantity responses to the crisis was obtained by examining the changes in agricultural export volumes and prices between the first quarters of 1997 and 1998. The results (shown in the table below) reveal a substantial increase in the volume of Thailand's crop exports since the crisis. The average increase of nearly 40 percent is the result of rises in the volume of exports of all commodities considered, except for tobacco. Some of these increases—for example, 76 percent for rice, 99 percent for fresh fruits, and over 200 percent for maize are remarkably large. The average volume increase is more than twice the volume growth of exports as a whole.

These increases in volumes were, however, offset by substantial declines in the dollar prices for most of these goods. The decline in the world prices of these same commodities can explain the decline in the prices of these exports. However, the declines in prices were particularly large relative to world prices for commodities, such as rice, fruit, and maize, in which Thailand greatly expanded its export volumes. Although these volume-induced increases are undoubtedly important, it appears that about 80 percent of the decline in the prices of these agricultural goods can be explained by developments in world prices.

The future performance of Thai agricultural exports will be of critical importance. Improved baht prices will undoubtedly increase production of agricultural commodities, leaving more crops available for exports. It seems unlikely, however, that the large improvement in agricultural prices in baht brought about by the devaluation will be sustained in the longer term.

Source: McKibbin and Martin, 1998.

Changes in Thailand's crop exports between the first quarter of 1997 and the first quarter of 1998

	Average share (Percent)	Quantity (Percent change)	Price (Percent change)	World price (Percent change)
Rice	36.3	75.9	-23.4	-14.1
Rubber	38.2	15.5	-39.3	-39.0
Tapioca products	12.9	8.3	-15.6	-10.7
Frozen fowl	6.9	49.7	-25.5	-8.3
Fresh fruits	0.7	99.0	-31.6	-7.0
Coffee	3.8	8.2	33.9	11.7
Tobacco leaves	0.7	-24.9	-23.0	-17.9
Maize	0.4	203.9	-57.3	-6.9
Total	100.0	39.7*	-26.7*	-21.8

* Weighted averages
Source: Bank of Thailand Monthly Bulletin; World Bank Commodity Data.

especially in critical scale and capital-intensive import-substitution sectors, not just in export-oriented platforms. Finally, governments have to continue to invest in education, especially secondary education, and labor upgrading to improve labor competitiveness. Regional coordination arrangements are working reasonably well thus far, but need to be more forward-looking in analysis and policy than they currently are or were during the crisis.

Notes

1. The results are based on the following panel estimations (fixed effects) for each country for the top ten manufactured exports of those countries (Bhattacharya, Ghosh, and Jansen, 1998)

$$\Delta S(i,t) = a + c(i) + d(t) + \beta_1 D_1(t) \Delta SC(i,t) + \beta_2 D_2(t) \Delta SC(i,t) + e(i,t)$$

where S denotes, for example, Malaysia's world market share, SC is the corresponding market share of China, Δ stands for the difference operator, and the indices i and t denote product category and time, respectively. The dummy variable $c(i)$ takes on a different value for each 2-digit SITC product group, whereas $d(t)$ takes on a different value for each period; $c(i)$ removes fixed differences between product groups, whereas $d(t)$ removes time-related factors common to all markets. D is a zero-one dummy variable that is used to divide the sample period 1989–96 into two subperiods: 1989–92 and 1993–96. The coefficients β_1 and β_2 measure the displacement effect of a Chinese market share increase of 1 percentage point in the two subperiods. The sample size is 80 observations.

2. Ernst, 1998.

3. Ernst, 1998.

4. Diwan and Hoekman, 1998.

5. Earlier studies have shown that if East Asia's export prices rise by 10 percent, their export quantities expand by 4 to 6 percent (see Dasgupta, Hulu and Das Gupta, 1995).

6. Import prices are also falling to the same or similar extent and therefore this does not necessarily cause any income terms of trade losses. Only when relative prices decline are there significant terms of trade deterioration.

7. Muscatelli, and others (1994).

chapter three

The Financial Sector: At the Center of the Crisis

Lim Beng Poon is suffering...Lim, a mid-level manager for a Malaysian bank, is feeling the pain of the ringgit's long slide against the U.S. dollar. It's hitting him where it hurts: his children's education. Lim has two college-age kids studying in the United States, and he pays their tuition in dollars. "My cost has gone up more than 15 percent," he says. "God, I hope this thing goes away." (Alkman Granitsas et al., "Now's Your Chance," Far Eastern Economic Review, *September 11, 1997) Indra Gunawan...plants himself in the throng massed outside the Jakarta headquarters of Bank Harapan Santosa. The bank is one of 16 shut down by the Indonesian government in early November in an effort to salvage the country's teetering financial system. The move has frozen the savings of thousands of people and cost 6,000 bank employees their livelihoods. Gunawan is one of them. 'I am here on behalf of my colleagues. They are just small employees—like drivers or office boys, but now they are suffering from losing their jobs,' he says, before policemen whisk him away, still wearing the [protest placards]... VICTIM OF LIQUIDATION around his neck."*—Suresh Unny, "The Next Battle," *Far Eastern Economic Review*, November 20, 1997.

For years, East Asia's "tigers" kept real deposit interest rates positive, progressively liberalized their financial systems, and worked to build a supporting institutional framework.[1] The financial sector deepened as these fundamentals encouraged household savings that were intermediated through domestic banking systems (see figure 3.1). This contrasted with other developing countries that experienced financial crises in which domestic financial intermediation was low; for example, in Mexico domestic credit to gross domestic product (GDP) was less than 50 percent in 1994, and in Argentina less than 20 percent in 1995. Despite the progress of East Asian financial systems, they developed without bond markets, lacked adequate prudential supervision, and in some countries permitted a large role for government. These weaknesses led to serious misallocation of resources, over-exposure to risky sectors, risky liability structures, and poor institutional development. Until recently, they were "covered up" by high growth, high savings, and strong fiscal positions.

While on a stock basis the region had, in 1996, among the highest ratios of equity market capitalization to GDP in the developed and developing world (see figure 3.1). In terms of new financial flows, in most East Asian countries, banking systems dominated financial intermediation. While equity markets developed rapidly in the 1990s, bond and other securities markets lagged. Where bond markets did develop, such as in the Republic of Korea, banks often guaranteed bond and other corporate securities, and secondary market trading was limited. Furthermore, rapid structural changes in the real economy were requiring improvements in the corporate governance of firms for which securities markets are important. This lack of well-developed capital markets meant that the monitoring of corporations was primarily the responsibility of banks and was not complemented by other financial institutions. The unbalanced financial systems also meant a lack of risk diversification, which amplified the shocks that hit the banking systems.

Even before 1997, the unbalanced financial systems were becoming a development challenge. In all countries in the region, non-bank financial markets were required to mobilize long-term funds to support the financing of infrastructure—with annual investment needs of up to US$150 billion—as well as other long-term investments, including housing. Existing pension systems in several countries were in need of reform, especially in the management of investment funds which, in turn, meant greater capital market development. While the population is relatively young in most East Asian countries, rapid aging was creating pressures to accelerate pension reform throughout the region. In contrast to trends in other developing regions, barriers

preventing foreign firms from entering the financial services industry in East Asia were high.

The problems in East Asia emerged, however, in the banking systems.[2] As many other developing countries, East Asian countries had some weaknesses in their banking systems including: low capital-adequacy ratios of banks; inadequately designed and weakly enforced legal lending limits on single borrowers or group of related borrowers; asset classification systems and pro-

FIGURE 3.1

Domestic credit provided by banking sector (1996)

Percent of GDP

Market capitalization of listed companies (1996)

Percent of GDP

Source: World Development Indicators, 1998 and IMF.

visioning rules for possible losses, which fell short of international standards; poor disclosure and transparency of bank operations; lack of provisions for an exit policy of troubled financial institutions; and weak supervision.

Weak governance of banks, often influenced directly or indirectly by government policies, added to the poor performance. Perhaps the most important weakness was the limited institutional development of banks. Much of the lending, for example, was done on a collateral basis, rather than on a cash-flow basis, thus obfuscating the need to analyze the profitability and riskiness of the underlying projects. Credit tended to flow to borrowers with relationships to government or private bank owners and to favored sectors, rather than on the basis of projected cash flows, realistic sensitivity analysis, and recoverable collateral values.

Responding to these deficiencies, East Asia was, in recent years, already placing increased emphasis on developing more efficient and competitive financial systems. Heavy state control of financial systems, which had worked well in some countries in the region, was being phased out in many countries. Countries were aware of the need to upgrade and remedy structural weaknesses in their financial systems and to liberalize in many dimensions. A Korean commission, for example, had recommended fundamental changes in its system in early 1997. The Philippines had been tightening the supervision of its banks. While difficult to quantify, it appears, however, that the improvements in the quality of supervision fell short of the rapid financial sector developments, particularly the speed of international financial integration. In other countries, often in response to financial crises, the quality of supervision has been strengthened in recent years—accompanied by increases in capital adequacy ratios, improved loan classification criteria, enlarged foreign ownership, higher liquidity requirements, and increased transparency. In East Asia, however, the urgency of reform was felt less.

Because banks appeared sound, there were few official or public demands on the government for greater reform. Data from individual annual commercial bank reports indicate that measured profitability during the 1990s was higher in many East Asian banks than in other countries. While overall intermediation costs, as

---

BOX 3.1

Offshore centers: The case of Thailand

The organizational design and special incentives of Thailand's Bangkok International Banking Facility (BIBF) contributed, to a great degree, to Thailand's crisis. In March 1993, permission was given to 46 domestic and foreign commercial banks to operate international banking business in Bangkok. In 1994, further privileges were granted to BIBF-based banks, including the right to open branches outside Bangkok and issue negotiable CDs. Due to special incentives, the BIBF provided an important channel for the domestic financial sector to raise short-term foreign-currency denominated funds ("out–in" lending). Because of bilateral tax treaties between Japan and Thailand, Japanese banks could offset withholding taxes levied on foreign exchange borrowings by Thai companies against their other income in Japan. As a result, Japanese banks, which had about one-quarter of the BIBF-market, were willing to absorb the withholding tax and lend at very low spreads to Thai companies. The supply of funds was further boosted by the incentive for foreign BIBFs to become full bank branches, the approval of which was made dependent on the volume of loans. Historically low international interest rates, especially on Japanese yen, were another factor in the large financing available and low spreads charged. As a result, out-in lending boomed between 1993 and 1996, from 194 billion baht to 794 billion baht. Reflecting the rapid growth of BIBF out-in lending, Thai commercial banks' foreign currency loans rose, at the end of 1996, to US$31.5 billion, or 17 percent of total private sector loans, while their short-term external liabilities surged.

Source: Kawai and Iwatsubo, 1998.

---

measured by interest margins (see figure 3.2), were high in East Asia,[3] costs to income ratios (see figure 3.3) did not suggest gross inefficiencies.

Nonetheless, segments of the financial systems in some of the East Asian countries were obviously performing poorly. State-owned banks in Indonesia, finance companies in Thailand, and merchant banks in Korea showed increasing signs of weakened performance and increased risks. Fragmented financial systems—(for example, commercial banks in Indonesia) and rules or other barriers prevented one class of financial services from competing against another (for example, banks and finance companies in Thailand)—further weakened competitiveness and stability, and added risk. In Thailand, for example, finance companies could not raise deposits; as a result they needed to rely on higher cost funds, which created incentives to lend for more risky projects. The large number of often weakly capitalized finance companies in Thailand and banks in Indonesia, for example, reduced the franchise value of financial institutions, thus undermining the incentives for prudent behavior. The competitiveness and institutional development of banking systems more generally was hampered by the low local presence of foreign banks.[4]

These weaknesses, whether considered individually or in total, were not sufficient to doom the financial system to immediate distress. All East Asian countries have performed well over the last decades in spite of these weaknesses. In the past, rapid economic growth and high domestic savings spared most East Asian banking systems from adverse consequences. Growing households savings, in the form of bank deposits, by capital inflows, and generally high enterprise profitability allowed banks to roll over non-performing loans and continue to carry bad assets on their books without experiencing liquidity problems. Also, strong fiscal positions enabled governments to reassure investors and depositors that they could back up their banking systems. Nevertheless, these financial sector weaknesses increased the risk of a financial crisis.

The process of financial liberalization unaccompanied by adequate supervision further added to these risks. Domestic and external financial liberalization led to increased competition for credit-worthy borrowers, reducing the franchise value of banks and induced them to pursue more risky investment strategies. Rapidly growing non-bank financial institutions (NBFIs), often lacking adequate internal, market and regulatory discipline, were an additional and important source of competition for banks, especially in Korea and Thailand. Furthermore, as NBFIs were generally less regulated and subject to weaker supervision than banks, their growth (in numbers and total credit expansion) directly exacerbated the system's fragility. Individual bank and non-bank data support this conclusion: finance companies in Thailand and merchant banks in Korea expanded rapidly and relied heavily on foreign exchange borrowings. State banks generally performed the worst and had the highest share of non-performing loans.

TABLE 3.1

Structure of the banking system before the crisis
(number of financial institutions)

	Private domestic commercial banks	State banks	Merchant banks	Finance or security co.	Foreign financial institution	Total
Indonesia	144	34[a]	0	0	44	222
Korea, Rep.of	26	8	30	53	52	169
Malaysia	23	1	12	40	14	90
Philippines	40	2	0	0	14	56[b]
Thailand	15	5	0	108	14	142

a. Includes 27 regional government banks.
b. Figures exclude thrift and rural banks.

Source: Bank staff calculations based on government sources.

The lingering effects of past policies to deal with financial distress exacerbated the impact of these weaknesses. Thailand (1983–87), Malaysia (1985–88), and Indonesia (1994) experienced financial crises that were, in part, resolved through partial or full public bailouts. These bailouts reinforced the perception of an implicit government guarantee on deposits, or even other bank liabilities, to the detriment of market discipline. In some cases, management of the restructured financial institutions was not changed, thus further undermining incentives for prudent behavior in the future.

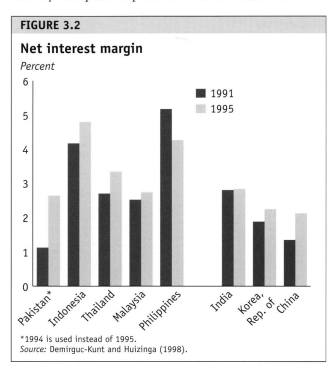

FIGURE 3.2

Net interest margin

Percent

*1994 is used instead of 1995.
Source: Demirguc-Kunt and Huizinga (1998).

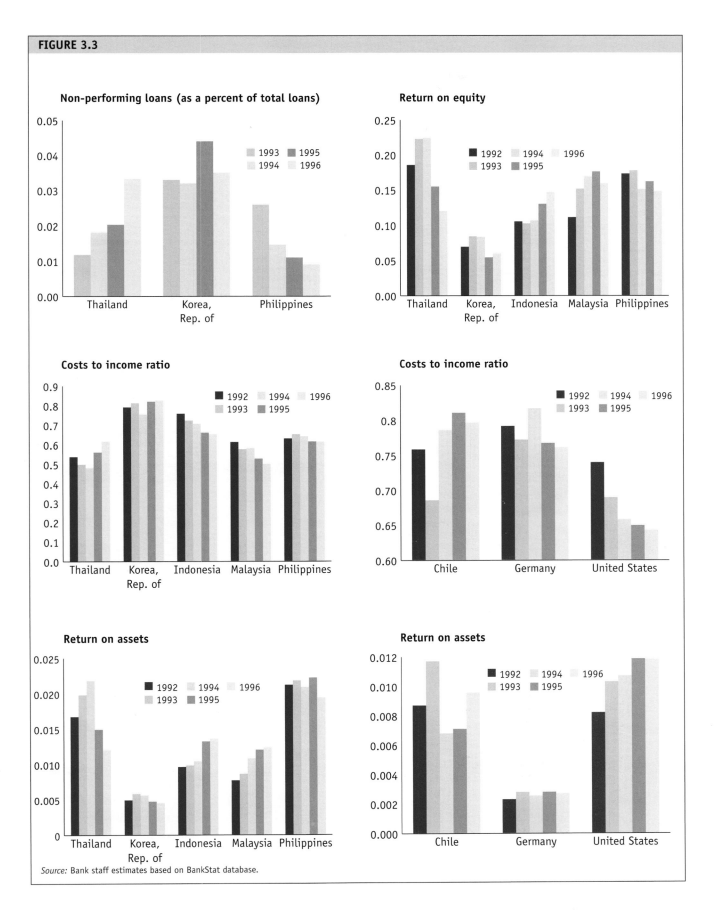

FIGURE 3.3

Non-performing loans (as a percent of total loans)

Return on equity

Costs to income ratio

Costs to income ratio

Return on assets

Return on assets

Source: Bank staff estimates based on BankStat database.

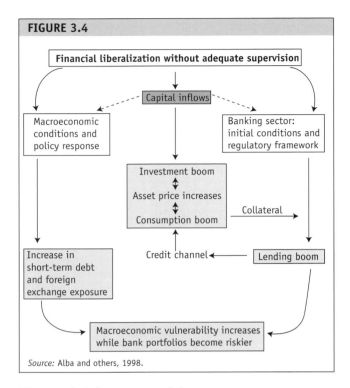

FIGURE 3.4

Financial liberalization without adequate supervision

Capital inflows

Macroeconomic conditions and policy response

Banking sector: initial conditions and regulatory framework

Investment boom
↕
Asset price increases
↕
Consumption boom

Collateral

Increase in short-term debt and foreign exchange exposure

Credit channel ← Lending boom

Macroeconomic vulnerability increases while bank portfolios become riskier

Source: Alba and others, 1998.

Financial boom and bust

As noted in chapter 1, the high growth rates in the region, partly the result of sustained structural reforms, led to demand pressures which were largely investment-driven. Although inflation increased in some countries, demand pressures were manifested primarily in a widening of current account deficits. Inflows of private capital contributed to, and reinforced, demand pressures. Private capital flows added to domestic macroeconomic pressures for three reasons. First, as countries had undertaken significant capital account liberalization, borrowing abroad could more easily accommodate demand pressures. Second, East Asian countries became more attractive to investors as their growth accelerated. Finally, asset price inflation already underway was fueled by private capital inflows. As a result, capital flows to East Asian countries tended to move very much in tandem with domestic macroeconomic cycles—particularly in Indonesia, Thailand, and Korea. The macroeconomic policy mix used to deal with the overheating pressures and capital inflows in the 1990s added impetus for further inflows—and for the accumulation of short-term, unhedged external liabilities, in particular. Exchange rate policies played a large role in motivating capital flows. The perception of reduced

exchange rate risk suppressed incentives to hedge external borrowing and moreover, created a bias toward short-term foreign currency borrowing.

Overall, the policy mix increased the incentives for unhedged external borrowing. During 1994–96, this policy mix was most pronounced in the cases of Indonesia and Thailand and, consequently, short-term liabilities rose very sharply in these two countries. The regulatory framework greatly influenced the differences in incentives and possibilities to borrow within countries. In Indonesia, for example, banks were limited in the amount of foreign exchange exposure they could assume and it was the corporate sector that borrowed large amounts offshore. In Korea and Thailand, NBFIs had relatively free rein to borrow abroad. In Malaysia, following large borrowings in the early 1980s, both financial institutions and corporations were limited in the degree to which they could take on foreign exchange exposures, thus foreign claims were relatively small in mid-1997.

Financial systems in the East Asia region generally did a poor job in intermediating capital. In fact, those countries, most notably Thailand, that relied most heavily on their banks, rather than foreign investment, were the most vulnerable.[5] As reviewed in chapter 1, implicit government guarantees, high domestic funding costs, and the creation of offshore banking centers all created incentives for excessive borrowing abroad.

Foreign investors were partially to blame for the over-borrowing, as they did not always adequately price risks, or perform full evaluations of countries or individual borrowers. Spreads for non-sovereign borrowers in East Asian countries, which were already lower than for other emerging markets, declined even faster relative to borrowers from other emerging markets during the 1990s, and in late 1996 and early 1997 were often only marginally above those for long-maturity loans to U.S. corporations. This decline in spreads was often accompanied by poor evaluations of risk and performance; as late as May 1997, for example, investors were buying large amounts of short-term paper from Indonesian corporations with only a few days of due diligence.

The reinforcing effects of high and rising investment levels, large private capital inflows and asset booms, combined with the underlying weaknesses in financial systems, led to the buildup of a number of vulnerabilities, including increased banking fragility, increased exposure to risky sectors, and increased borrowing short in foreign currency and lending long in domestic currency. As seen in figure 3.4, these effects not only reinforced macroeconomic policy, but they reinforced each other. There were, however, differences between countries that played important roles in both triggering the crisis and in the evolution of the crisis once it was under way.

Throughout East Asia, financial liberalization, increases in financial savings, and surges in capital inflows led to increases in monetary aggregates. In turn, as governments failed to make the judgment that monetary and credit growth, as well as capital inflows, were excessive, the increased liquidity and monetization of these economies resulted in a generalized surge in bank and NBFI lending, although the amplitude and duration of these cycles, plus their apparent relationship with financial liberalization and the surge in capital inflows, varied from country to country. For example, in Malaysia, the Philippines, and Thailand, bank and non-bank credit to the private sector began growing at higher rates and on a sustained basis fol-

lowing the surge in capital inflows. This high growth rate strained banks' capacity to adequately assess risk. In Indonesia, in contrast, the growth in bank and non-bank credit to the private sector was lower during the inflow period than in the years prior to the surge in foreign capital, in part as it has undertaken extensive financial reform before the inflow period (see figure 3.5).

The increasing financial sector fragility was not detected during the lending boom because the growth in banks' loan portfolios was accompanied by rising measured profits. Figures 3.3 and 3.5 show that in countries with high credit growth—except the Philippines—measured profitability of the banking sector increased consistently across all indicators. Conversely, in countries where the lending boom was smaller—in absolute terms or proportional to GDP—profitability increases were smaller or non-existent, depending on the profitability indicator used.

Real estate lending was high, and banking sector exposure to real estate was greater in countries with larger rates of credit growth proportional to GDP

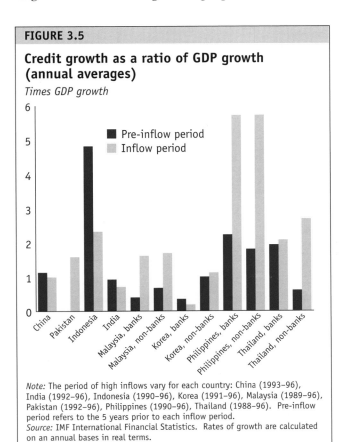

FIGURE 3.5

Credit growth as a ratio of GDP growth (annual averages)

Times GDP growth

Note: The period of high inflows vary for each country: China (1993–96), India (1992–96), Indonesia (1990–96), Korea (1991–96), Malaysia (1989–96), Pakistan (1992–96), Philippines (1990–96), Thailand (1988–96). Pre-inflow period refers to the 5 years prior to each inflow period.
Source: IMF International Financial Statistics. Rates of growth are calculated on an annual bases in real terms.

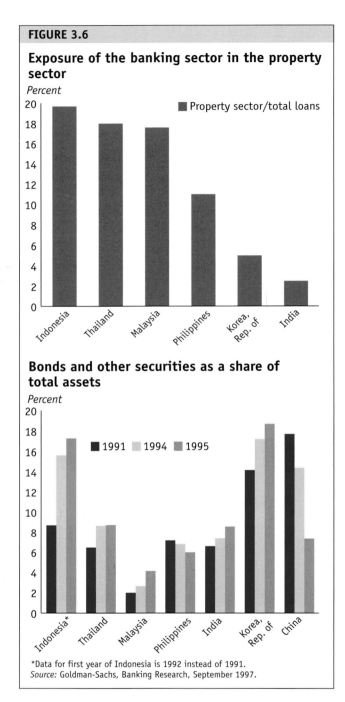

FIGURE 3.6

Exposure of the banking sector in the property sector

Percent

■ Property sector/total loans

(Indonesia, Thailand, Malaysia, Philippines, Korea, Rep. of, India)

Bonds and other securities as a share of total assets

Percent

■ 1991 ■ 1994 ■ 1995

(Indonesia*, Thailand, Malaysia, Philippines, India, Korea, Rep. of, China)

*Data for first year of Indonesia is 1992 instead of 1991.
Source: Goldman-Sachs, Banking Research, September 1997.

Indonesia, the variability of real estate prices was lower, with high to low ratios of 1.25 and 1.32, respectively. However, vacancy rates in 1996 were relatively high, at around 14 percent and increasing. In the Southeast Asian countries, especially Thailand, the amount of construction at the end of 1996 already suggested a significant oversupply of real estate for the following two years (1997–99).

But, differences among countries were significant. Korean banks, for example, did not have large property exposure. Korean banks did, however, increase the share of bonds and other securities in their portfolios to almost 20 percent (in addition, Korean banks extended large amounts of guarantees to securities issued by corporations). Except for the Philippines, countries generally increased their exposure to bonds and other securities.

In Thailand, Malaysia, and the Philippines, banks' foreign exchange exposures increased significantly since the late 1980s (see figure 3.7). Also, for Thailand and the Philippines, the stock of foreign liabilities of NBFIs increased rapidly. In Indonesia, the increase of banks' foreign exchange exposure was significant through 1994, when there was a slight decrease. During the same period, commercial banks in Korea did not show a large increase in foreign exchange exposure, unlike Korean merchant banks and the corporate sector both of which increased foreign exchange exposures significantly.

Maturity mismatches created another vulnerability, especially on the external financing side. Initial levels of external debt in East Asia were low by international standards (see figure 3.7), with the exception of Indonesia. However, during 1991–96, short-term external liabilities accumulated rapidly (see figure 3.9) and most of this borrowing was unhedged. In Thailand, the Philippines, Korea, and Malaysia, short-term foreign liabilities of banks grew extremely rapidly. Indonesian banks' short-term foreign liabilities did not increase rapidly, but that of Indonesian corporations did. The crisis itself has revealed that short-term borrowings were even higher than these figures suggest because much non-bank liabilities and borrowing escaped national and international (BIS) coverage.

The strongest indicator of vulnerability was the ratio of short-term external debt to external reserves prior to the crisis (see figure 3.10). In June 1997, short-term

growth (see figure 3.6).[6] This created risks, as both the real estate and securities markets have been very volatile in East Asia. Real estate price fluctuations during the 1990s were the highest in the Philippines and Malaysia, with a ratio of highest to lowest prices since inflows started of 3 and 2, respectively. Still, in both countries, vacancy rates in 1996 were relatively low at around 2 percent (and the banking sector exposure to real estate appeared to be low in the Philippines). In Thailand and

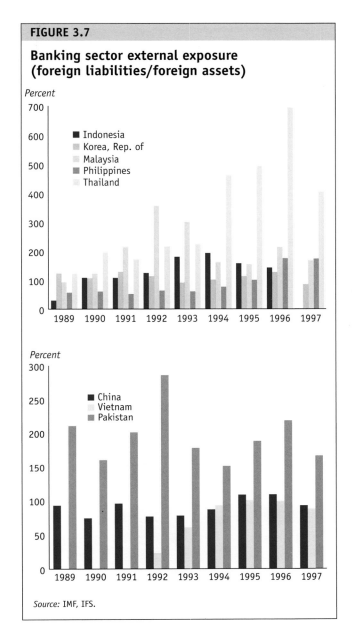

FIGURE 3.7

**Banking sector external exposure
(foreign liabilities/foreign assets)**

Percent

- Indonesia
- Korea, Rep. of
- Malaysia
- Philippines
- Thailand

1989 1990 1991 1992 1993 1994 1995 1996 1997

Percent

- China
- Vietnam
- Pakistan

1989 1990 1991 1992 1993 1994 1995 1996 1997

Source: IMF, IFS.

inflows of private capital. However, the expansion of international integration, and easy access to private capital flows became catalysts which increased the magnitude of vulnerability. Nonetheless, there were important differences among countries. Malaysia and the Philippines were less exposed; in Thailand, finance companies and real estate lending created large risks; in Korea, merchant banks were the main source of short-term foreign exchange claims; and in Indonesia, corporate lending was large.

The differences in vulnerability between countries were influenced significantly by the sequencing and process of financial liberalization. In Korea, for example, corporations were limited in the degree to which they could raise funds abroad or sell domestic-currency denominated securities to foreign investors. Merchant banks, however, had *carte blanche* to borrow abroad, and were subject to only limited supervision. Many of the merchant banks were under *de facto* control of the chaebols, and were able to channel foreign funding to them.

The fact that very few observers predicted the crisis corroborates the view that the source of the crisis lies with the vulnerability of financial structures, importantly, rather than with fundamental weaknesses in the financial sector alone (see box 3.3).

From bust to reconstruction: Moving forward and rebuilding

Improving the regulatory framework

Since the onset of the crisis, countries in the region have begun to improve the regulatory and supervisory framework for financial institutions. They have recognized the need to improve standards for capital adequacy, loan classifications, and loan provisioning in order to move closer to international practice. Not all of these requirements are immediate, and some will be phased in over time to give banks time to adjust and build up their capital. Some countries have enhanced disclosure requirements, increased the (legal) responsibilities and duties of managers of financial institutions, and promoted greater roles for outside investors. Countries have made progress in putting in place the tools for financial sector restructuring, including the

debt in Korea, Thailand, and Indonesia exceeded external reserves by a large margin—greater than that of many developing countries. This high ratio of short-term obligations to liquid foreign exchange assets rendered these countries much more vulnerable to a potential run on their currencies, which could arise from a loss of investor confidence.

Riskier investments, together with the banking sector's growing fragility and the accumulation of short-term external liabilities, culminated in increasing macro and financial vulnerability. The underlying process of vulnerabilities may have occurred even without the

TABLE 3.2
Indicators on accounting and prudential standards

	Non-performing loans (NDL)definition (Number of months overdue)	General provision (percent loans)	Loss provision (percent of NPL)*	Capital-asset ratio (percent)
Malaysia	3	1.5	1.5, 20, 50, 100	8 now; 10 by year 1999
Indonesia	3 by year 2001	1.0	5, 15, 50, 100	4 now; 12 by year 2001
Korea, Rep. of	3	0.5	2, 25, 75, 100	8
Philippines	3	1.0 (raised to 2.0 by Oct. '99)	2, 25, 50, 100	10
Thailand	3 by year 2000	1.0	2, 25, 50, 100	8.5

*Special mention, substandard, doubtful, loss provision standards.
Source: World Bank staff.

development of the institutional and legal framework for resolving distressed financial institutions and dealing with non-performing assets.

TABLE 3.3
Foreign ownership limits for existing institutions

Indonesia	No limits
Malaysia	30% of total equity
Philippines	60% of total equity
Korea, Rep. of	10% of total equity; 25%-33% with special permission
Thailand	100% for 10 years

Source: Various government publications.

All countries have opened up their markets to foreign investors to greater degrees through more liberal rules for purchases of domestic assets (both capital markets instruments and foreign direct investment) and by easing local establishment rules. Korea, for example, has completely liberalized foreign purchases of domestic bonds and has abolished most previous ownership restrictions for foreign investors. Thailand now allows foreign ownership of banks and corporations (albeit with some limits), and Indonesia has, with a few exceptions, liberalized foreign ownership of all types of financial institutions and corporations.

Restructuring the financial sector

Most important, countries have implemented some financial sector restructuring measures (see table 3.4). Thailand has closed more than half of its 91 finance companies, while Korea has closed about half of its merchant banks. Indonesia has set in place a process for dealing with insolvent banks: It has closed 16 financial institutions and has put another 54—including several of it largest banks—under the control of its bank

restructuring agency. The government of Korea has taken control of two insolvent commercial banks, recapitalized them, and expects to sell them to foreign investors in the near future. It also shut down 16 of its 30 merchant banks. The governments of Korea and Indonesia have provided support to commercial banks by taking over distressed loans and providing balance sheet support. In Thailand, several commercial banks have received some infusions of new capital through new issues and as many as four Thai commercial banks have had some foreign investor equity participation. Malaysia is in the process of restructuring its finance companies' industry through mergers and closures—involving 31 finance companies—and has created an

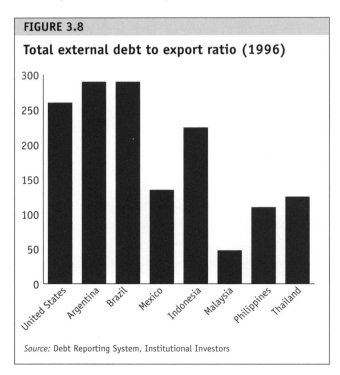

FIGURE 3.8
Total external debt to export ratio (1996)

Source: Debt Reporting System, Institutional Investors

BOX 3.3

Did anyone foresee the crisis?

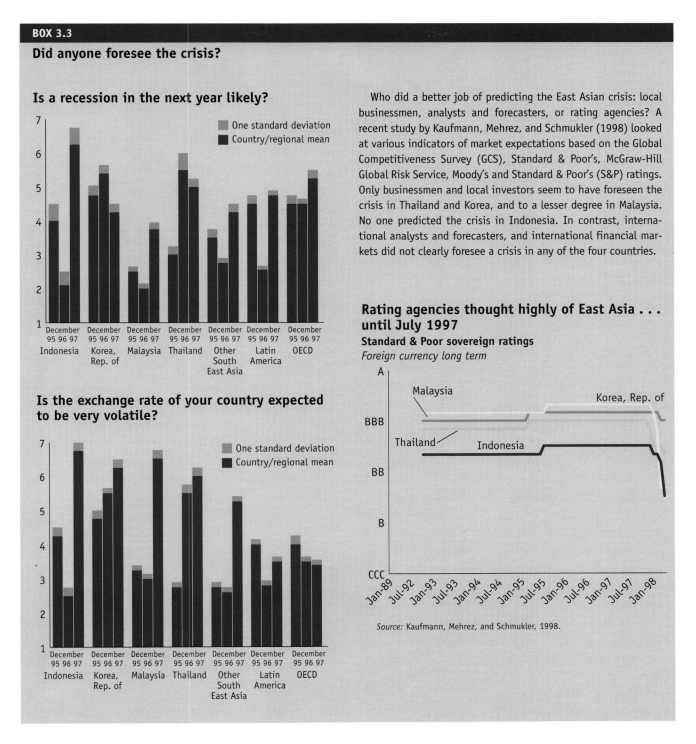

Is a recession in the next year likely?

One standard deviation
Country/regional mean

| December 95 96 97 | December 95 96 97 | December 95 96 97 | December 95 96 97 | December 95 96 97 | December 95 96 97 | December 95 96 97 |
| Indonesia | Korea, Rep. of | Malaysia | Thailand | Other South East Asia | Latin America | OECD |

Is the exchange rate of your country expected to be very volatile?

One standard deviation
Country/regional mean

| December 95 96 97 | December 95 96 97 | December 95 96 97 | December 95 96 97 | December 95 96 97 | December 95 96 97 | December 95 96 97 |
| Indonesia | Korea, Rep. of | Malaysia | Thailand | Other South East Asia | Latin America | OECD |

Who did a better job of predicting the East Asian crisis: local businessmen, analysts and forecasters, or rating agencies? A recent study by Kaufmann, Mehrez, and Schmukler (1998) looked at various indicators of market expectations based on the Global Competitiveness Survey (GCS), Standard & Poor's, McGraw-Hill Global Risk Service, Moody's and Standard & Poor's (S&P) ratings. Only businessmen and local investors seem to have foreseen the crisis in Thailand and Korea, and to a lesser degree in Malaysia. No one predicted the crisis in Indonesia. In contrast, international analysts and forecasters, and international financial markets did not clearly foresee a crisis in any of the four countries.

Rating agencies thought highly of East Asia . . . until July 1997

Standard & Poor sovereign ratings
Foreign currency long term

Malaysia
Korea, Rep. of
Thailand
Indonesia

Source: Kaufmann, Mehrez, and Schmukler, 1998.

agency that will take over non-performing assets from banks and finance companies.

On the other hand, countries have also limited the number of options for dealing with distressed financial institutions.[7] Most importantly, in attempts to restore confidence, among both domestic and foreign investors, most governments have issued statements on

the solvency of their financial systems. Thailand, Korea, Indonesia, and Malaysia have all issued formal guarantees. The Indonesian guarantee is the most comprehensive as it covers all deposits and creditors, both in rupiah and foreign currency and for on- as well as off-balance sheet liabilities, with, among other things, equity and subordinated debt not covered. In the case

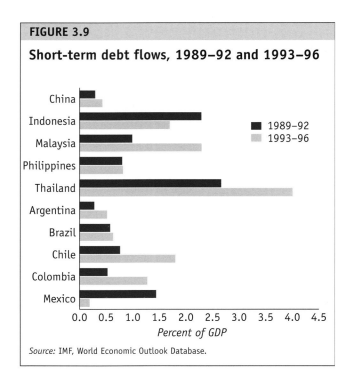

FIGURE 3.9

Short-term debt flows, 1989–92 and 1993–96

Percent of GDP

Source: IMF, World Economic Outlook Database.

of Korea, the government has guaranteed a substantial amount of the external debt incurred by merchant banks and other financial institutions; these claims have been restructured into loans with higher spreads and maturities up to 10 years. As a result of these guarantees, governments have limited the degree to which private creditors of financial institutions can be asked to share the burden of financial restructuring. Governments have become the *de facto* owners of many insolvent financial institutions, which has become an important factor in determining the scope of financial restructuring options. Most countries are planning to phase out these guarantees in lieu of more limited, formal deposit insurance schemes, but this will be gradual in order to maintain general confidence in the financial system.

Responses elsewhere in East Asia

The East Asian financial crisis has led to intensified global efforts to develop more robust financial systems in emerging markets and developing countries. China is accelerating the process of enterprise restructuring and financial sector reform. Many unsound state-owned enterprises (SOE) have been closed and their assets disposed of. Resolution Trust Company-type operations

are furthering the highly decentralized SOE reform and privatization tasks that lie ahead. The government has announced that it will provide US$32 billion to its state-banks to raise capital adequacy as to step toward complying with Basle standards. This is a small amount relative to the reported size of non-performing assets in China's financial system, and more fiscal resources are likely to be required. The government may establish one or more Asset Management Corporations to help dispose of state-owned commercial bank assets associated with non-performing SOE loans. However, the Chinese government faces the dilemma of the competing priorities of preserving social stability versus restoring financial sector health. The pace of reform will be determined by how these two are balanced. Solid economic growth requires gradual, but firm, reform.

To date, very few East Asian corporations have been restructured. All East Asian countries have recently tried to facilitate enterprise restructuring by creating an enabling environment, which includes better accounting and disclosure standards, bankruptcy and foreclosure processes, and taxation and accounting rules. Indonesia adopted a new bankruptcy code in August 1998, Thailand revised its bankruptcy law in March 1998, and Korea introduced some revisions to its corporate reorganization procedures in February 1998. But workout mechanisms are relatively poor and untested, and local skills in restructuring are limited and inefficient. In many East Asian countries foreclosure of collateral is particularly weak—in the past it often took several years. Large-scale technical assistance is being provided, including foreign auditors and consultants to evaluate asset portfolios and financial institution rehabilitation plans.

TABLE 3.4

Indicators on consolidation progress

	Initial financial inst.	Closed or suspended	Nationalized or under supervision	To be merged	Bought by foreigners, joint venture
Indonesia	222	16	54	4	0
Korea, Rep. of	169	16	2	5	0
Malaysia	90	0	4	31	0
Thailand	142	53	18	0	4
Philippines	56[a]	2	0	0	0

Note: a. Excludes thrift and rural banks.

Source: World Bank staff based on government publications.

Countries have encouraged capital market development by removing regulatory and tax impediments and enhancing the institutional framework for primary and secondary markets. The need to raise public resources for the restructuring of the banking system has provided the impetus to develop bond markets. The governments of Korea and Thailand have announced they will issue bonds equal to about 15 percent of their respective GDPs. Initially, these bonds will be placed with financial institutions to strengthen their income and asset portfolios, and secondary market trading is likely to be limited. Over time it is expected that government bonds will become a benchmark for corporate bonds, thus fulfilling a function that was previously missing when public domestic debt was minimal. In addition, countries like Korea will standardize and rationalize the issuance of government bonds. As a result of these actions, governments have built a base for more balanced financial systems which rely less on bank financing, more adequately diversify risk, and allow for better monitoring of corporations. But many impediments remain, most notably for corporate bonds.

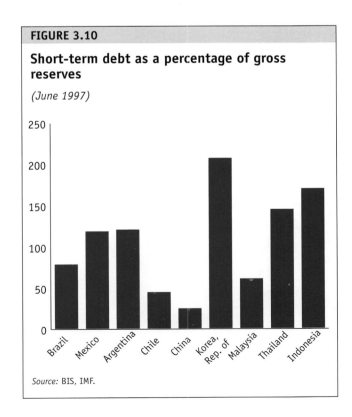

FIGURE 3.10

Short-term debt as a percentage of gross reserves

(June 1997)

Source: BIS, IMF.

The short-run agenda: Restoring credit flows

The growth rates of loans, adjusted for inflation, in Indonesia, Korea, and Malaysia has declined sharply since the crisis (see figure 3.11). This is consistent with firms' complaints, especially in early 1998, about the lack of credit to finance their production and exports. The growth rate of real loans in Thailand is an anomaly, and may be explained as banks extending lines of credit to existing clients, although they face financial difficulty, in order to prevent a further increase in the volume of non-performing loans. The evolution of aggregate loans should be interpreted with caution when assessing credit conditions. When the economy is hit by a negative shock, it is often impossible to distinguish whether the usual deceleration in bank lending stems from a shift in demand or rather in supply. On the one hand, the corporate sector may be demanding less credit because they are undertaking fewer investments; on the other hand, it could be that banks are less willing to lend and, therefore, charge higher interest rates or decline more credit applications. Analysis of credit aggregates and relevant spreads indicate that credit conditions were indeed tight in Indonesia, Korea, and Malaysia in the first quarter of 1998 and less so for Thailand and the Philippines (see box 3.4).

Because of large uncertainty, many banks have preferred to cut back on lending and invest excess liquidity in less risky assets, such as government securities. As aggregate domestic demand has contracted and exports have not yet increased in value terms, the rate of increase in demand for credit has fallen. Also, corporate balance sheets and profitability have deteriorated, and collapsing asset values have cut collateral values. As a result, lenders are less willing to extend new credit. Banks are even more reluctant to lend; their financial positions have weakened due to losses on foreign exchange and deteriorating asset portfolios, and prudential regulations have been tightened.

These micro-factors behind a credit crunch also predominated in many other episodes. In the mid- to late 1980s, the balance sheets of industrial countries' commercial banks were under pressure from exposure to developing countries with debt-servicing difficulties and real estate downturns. Their credit crunch was only

FIGURE 3.11

Growth rate of real loans

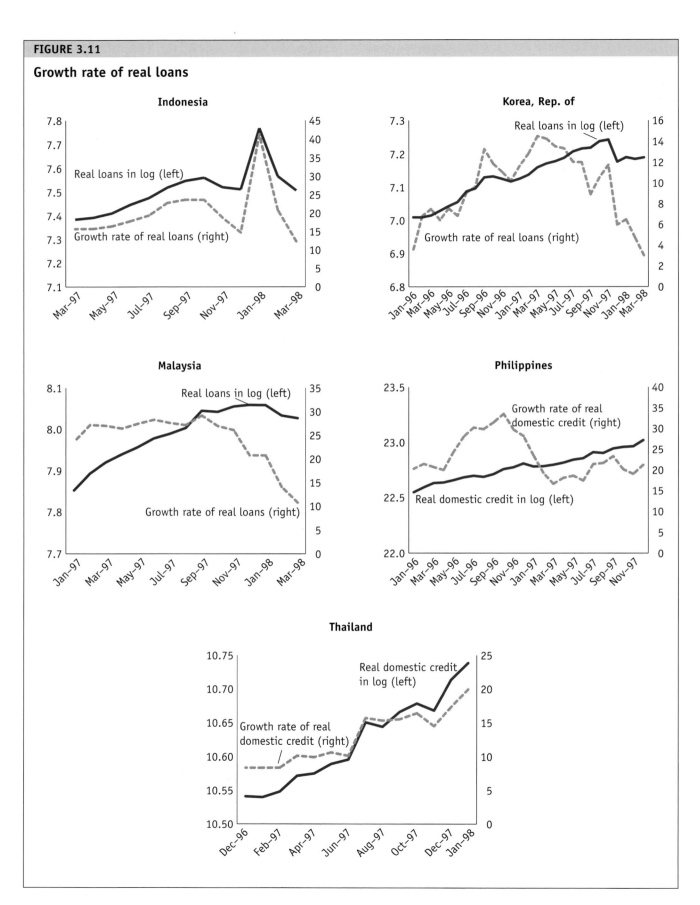

BOX 3.4

The credit crunch in East Asia

The issue of the credit crunch in the aftermath of the Asian crisis has stimulated much debate. Indeed, some of the features of the East Asian economies, for example, bank-based financial systems and high leverage, make them particularly vulnerable to monetary/financial shocks. Under these circumstances, the credit channel of transmission of these shocks is likely to lead to a credit crunch. In particular, the credit channel holds that a monetary/financial shock adversely affects the flow of bank loans to agents—e.g. households and small and medium-sized Enterprises (SMEs)—for whom close substitutes for bank credit are unavailable. In turn, the disruption of the availability of finance for bank-dependent borrowers may hinder economic activity.

In practice, however, it is difficult to detect the credit channel effects that lead to a credit crunch. Reliance on trends of credit aggregates alone is inadequate to prove that there has been an adverse shift in the supply of loans: even a decline or slowdown of credit could stem from a decrease or deceleration in demand. A methodology frequently used to overcome this focuses on both credit aggregates and the yield spread between bank loans and risk free assets—for example, government bonds. If this spread rises while credit aggregates slow down, it can be conjectured that the supply of loans has either decreased more or increased less than demand. However, further qualification is needed. The increase of the spread might simply reflect a rising risk premium triggered by the fact that the negative shock reduces the net worth of economic agents due to larger financial outlays. Accordingly, the relevant spread to capture the worsening credit conditions that affect bank-dependent borrowers is the spread between bank lending rates and corporate bonds. The yield spread between corporate and government bonds measures the general risk premium.

Ding, Domaç, and Ferri (1998) apply this methodology to crisis countries and show that, in the first few months after the crisis, the credit crunch is widespread, while its negative impact particularly affects small-sized banks and enterprises.

Along with a slowdown in credit aggregates, the evolution of the two spreads for Korea underscores that both the general risk premium and the bank-dependent borrowers' specific spread have increased markedly following the sudden depreciation of the won and the sharp increase in interest rates. In Korea, such increases in the spreads have been found to systematically lead to a decline in industrial production (Domaç and Ferri, 1998).

Interest rate spreads for Korea

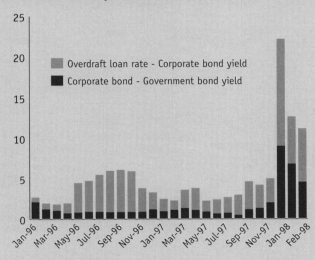

Source: Ding, Domaç and Ferri, 1998.

alleviated when exposure had been resolved. This was accomplished in part through the Brady plan for developing countries' debt, the recovery in prime markets, and the buildup of capital from retained earnings. In Japan, the effects on the real economy of current problems in the financial sector have a similar pattern of micro-factors, as monetary conditions in Japan are relaxed and interest rates are historically low. Therefore, there must be improved structural and sectoral policies in the financial sector to fully restore credit flows.

Nevertheless there are some useful short-term interventions that will activate credit flows. First, enhanced security needs to be provided for those creditors willing to advance new money to debtors that are operationally viable but financially overextended with the concurrence of existing creditors. This type of "debtor-in-possession," or "DIP" financing is standard practice in market economies for firms in reorganization. It could be done on an emergency basis and for a large share of firms. However, caution is necessary because in these economies the problem is much larger and the institutional capacity for corporate reorganization is much weaker, and repayment discipline on old debts needs to be maintained for those borrowers able to pay.

In some countries, this policy will only be effective if the government can move quickly to select viable entities—banks and corporations. In Indonesia, for exam-

BOX 3.5

The fiscal costs of bank restructuring: Three scenarios

The number of variables involved, and the current uncertainties on the policy choices to be made allow only for some illustrative scenarios for the fiscal costs of bank restructuring, presented in the table below. The scenarios illustrate the net fiscal costs, that is net of the recovery on bad assets taken over by government. All three scenarios assume that (a) one-third of the bad debts transferred to government in the course of recapitalization will be recovered and (b) the bad debts not taken over by the government can be written off against future profits. This adds to the fiscal costs at the rate of the tax rate times the remaining stock of bad debts. The annual fiscal costs will be the interest burden on the stock amount of public resources needed.

Low Scenario. The low scenario assumes a low share of bad debt and no recapitalization. The only fiscal costs are the debt write-offs against profits.

Medium Scenario. The medium scenario assumes a low share of bad debts, but recapitalization up to the point at which the banks have non-negative capital. Any remaining bad debt after recapitalization is written off against profits.

High Scenario. The high scenario assumes a high share of bad debt, and recapitalization up to 4 percent of remaining assets. Any remaining bad debt is written off against profits.

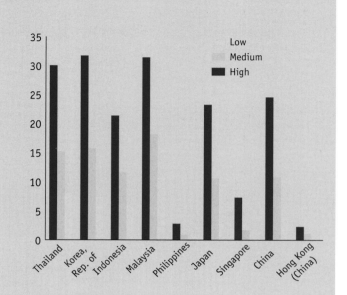

Fiscal costs of bank recapitalization (principal amounts)

Bad Debt as a percentage of financial assets

	Thailand	Korea, Rep. of	Indonesia	Malaysia	Philippines	Japan	Singapore	China	Hong Kong (China)
Low and Medium	20.0	20.0	30.0	20.0	10.0	10.0	5.0	20.0	5.0
High	35.0	35.0	40.0	30.0	15.0	20.0	22.0	40.0	10.0

Source: World Bank staff estimates, Demirguc-Kunt and Huizinga (1998); Deutsche Bank, JPMorgan, IFS.

---

ple, the resumption of credit flows will require mechanisms to quickly offer support to some of the institutionally better-developed banks, and through those banks, to corporations. Even though some banks now have weak asset portfolios, there are few alternatives to supporting some of them immediately; institutionally strong banks should form the core of a restructured banking system.

Also, specialized financing facilities might prove useful. In trade financing, Indonesia, Korea, and Thailand have already established public support programs—the central banks purchase notes collateralized by export receipts, while government supports the commercial banks to issue letters of credit. Some countries may require further government support for trade financing and working capital. This support could be in the form of a guarantee on the performance of the better banks in the respective country rather than a direct and full credit guarantee on particular loans or credits. In turn, this sovereign guarantee might be supported by third

parties, such as in the case in Indonesia and Thailand, which have backup guarantees from the Asian Development Bank (ADB) and other bilateral lenders.

In the short-term, further measures directly geared toward the corporate sector, which boost cash flow, and thus boost credit standing, are likely to be the most effective. Because corporations are relatively strong in East Asia, the measures should be self-selective and in order to avoid the adverse selection often associated with government-directed schemes. Such schemes may include tax relief, credit for collateralized transactions, and cash flow enhancement through wage moderation. Support has also already come through some tax measures, for example, Thailand has eliminated income taxes on accrued unpaid interest. The careful design of these short-term measures will be necessary and they should not become permanent features of the landscape. And, fully resolving the credit crunch will require bank restructuring programs.

The difficult and costly task of bank restructuring

Systemic bank distress is prevalent in East Asian countries. The process of closing banks and other financial institutions and restructuring the banking system will be difficult and costly. Private sector estimates show ratios of non-performing to total loans for East Asian countries (excluding Hong Kong (China) and Singapore) between 10 and 60 percent, with implied ratios of non-performing loans to GDP between 10 and 40 percent. Official figures lag these estimates, but some countries already have estimated ratios of non-performing loans to total loans above 20 percent. To include the distressed assets held by other financial institutions, such as insurance companies and investment funds, would raise these estimates of non-performing assets even further, especially in Korea and Indonesia. Banks' capital adequacy positions are also very weak; capital to asset ratios after write-offs of non-performing loans are estimated to be between -17 percent (Indonesia) and -4 percent (Malaysia); banks in Singapore, Hong Kong (China), and the Philippines vary from 10 to 16 percent.

Since governments have provided guarantees for the liabilities of many financial institutions, and it is also in the national interest to resolve these losses as quickly as possible, most costs of the financial restructuring will be a fiscal liability. While some costs will be borne by the private sector, estimates imply costs for the government of up to 30 percent of GDP. This is very high even relative to international experience. Chile's 1982 banking crisis was the most expensive known to date, and the total fiscal cost was 41 percent of GDP—not much higher than current private sector estimates for some East Asian countries. However, a large up-front public sector investment may lead to lower ultimate costs by avoiding the moral hazard of repeated bailouts and by reaping the large benefits of activating credit flows. But, large resources require a greater government capacity to raise and service non-inflationary funds and improve fiscal management. In some countries, most notably Indonesia, public debt may not allow this.

Principles of bank restructuring

In East Asia the challenge is to maintain confidence while dealing with insolvent financial institutions. Many of the banks in the region also require capital and must be strengthened to restore credit flows. The target for rebuilding capital in East Asian banks generally has been set at 8 percent or higher. The time table for meeting these targets varies by country, but bank supervisors have generally set a 1- to 2-year deadline. Banks unable to meet that deadline on their own will require outside support.

International experience offers some principles for bank restructuring. First, only viable institutions should stay in business, and restructuring should allocate losses transparently while minimizing the cost to taxpayers. Second, restructuring should strengthen financial discipline by allocating losses first to existing shareholders; and then to creditors, and perhaps large depositors. Third, measures should be taken to maintain credit discipline for bank borrowers and to preserve incentives for the infusion of new private capital. Fourth, restructuring should be fast enough to restore credit, while maintaining confidence in the banking system. Given these principles, governments could evaluate the various policy options at hand. Table 3.5 provides an overview of options for bank restructuring.

TABLE 3.5

Different options for bank restructuring entail trade-offs

	Speed	Fiscal costs	Incentives for bank performance	Confidence in banking system
Bailout	●●●●●●	●	●	●●
Assisted mergers	●●●	●●●●●	●●●	●●●●
Recapitalization and sale	●●	●●●●	●●●●●	●●●●
Restructuring plan	●●	●●●●●●	●●●●	●●●
Liquidation and payoff	●●●●●	●●●	●●●●●●	●

Note: More dots should be interpreted as "better."
Source: World Bank staff evaluation.

Selection of each option involves trade-offs (see table 3.5). For example, if the government were to "bail out" troubled banks by supplementing bank capital with public resources, the strategy would have the advantage of speed, but at a high cost to the treasury and low incentives for bank managers to improve performance. On the other hand, if the government were to recapitalize distressed banks and sell them later, this would

lower fiscal costs, provide better incentives for bank performance, and raise confidence in the banking system, but at the probable cost of making the process longer. Liquidating and paying off creditors and depositors would provide a speedy solution, give better incentives for bank performance, and entail relatively low fiscal costs, but this might severely undermine the confidence in banking system. Balancing these trade-offs will vary from country to country, and depend on the nature of the systemic problem.

Approaches taken to date

Financial restructuring determines not only the allocation of current losses, but also the distribution of ownership and future control in the economy. Each country is going through a process of political and economic decision making to resolve these issues. In Korea, the government has adopted a mix of a decentralized and centralized solution; banks have been partially recapitalized and some non-performing loans have been transferred to the government-owned Korean Asset Management Company (KAMCO), which will restructure these loans. Thailand created the Financial Restructuring Agency to oversee rehabilitation of finance companies and commercial banks. The Thai government-owned Asset Management Company, in competition with the private sector, can acquire assets from defunct finance companies and then sell them over time, including through auctions, while banks have been partly maintained through liquidity support from the Bank of Thailand. In Indonesia, the IBRA has taken over 54 banks and initiated restructuring of their portfolios. Thailand has progressed the furthest with financial restructuring, especially the finance companies' sector, but even there the task is not nearly complete.

The restructuring will require a deliberate policy of picking winners among banks and corporations. In Indonesia, for example, the resumption of credit for working capital and trade financing will depend on whether mechanisms can be found to support some of the institutionally better-developed banks, and thereby support corporations. Currently, these banks may have weak asset portfolios but the institutionally strong banks will have to form the core of a restructured banking system—short of shutting down the whole banking

---

BOX 3.6

World Bank efforts in the financial sector

The World Bank has been working closely with governments in crisis countries to restructure their financial systems. In Thailand, the US$15 million Financial Sector Implementation Assistance fast-track loan of September 1997 was intended to enhance the resiliency and soundness of the financial sector. This first loan was followed quickly by the December 1997 US$350 million Finance Companies Restructuring Loan, the February 1998 US$15 million Economic Management Implementation Assistance Loan, and the Economic and Financial Adjustment Loan of June 1998. The World Bank has helped to establish a financial restructuring agency to manage the reforms of the suspended finance companies and an asset management company to recover assets.

In Indonesia, a US$20 million Banking Reform Assistance Loan has assisted the newly-created Indonesian Bank Restructuring Agency in resolving the problem of distressed finance companies. The Bank has also supported the government's efforts to restructure private corporate debt. The July 1998 Policy Reform Support Loan commits US$1 billion with the objectives of stabilizing the banking system, bolstering depositor confidence, and ensuring continued essential banking services in the economy. The US$3 billion Economic Reconstruction Loan (ERL) to Korea in December 1997 helped stabilize the currency and support efforts to restore the financial system. The government's program suspended insolvent financial institutions, helped to restore capital adequacy, improved supervision, halted all direct lending, reinforced the deposit insurance function, and improved the efficiency of financial entities. Building on the ERL, the US$2 billion Structural Adjustment Loan in March 1998 supported financial sector restructuring and development, corporate sector reform, and capital market development. The World Bank is working with the government of Korea to help develop an efficient, diversified, and competitive financial system, including insurance and securities industries.

For Malaysia, one of the aims of the World Bank's Economic Recovery and Social Sector Loan of June 1998 was to strengthen the financial sector through consolidation of finance companies via assisted mergers, preemptive recapitalization of banking institutions, improvement in investor protection and reduction in settlement risks, and by progressively liberalizing the level of foreign ownership. If these programs are implemented vigorously, they will contribute to improving confidence in capital markets, and transparency.

Source: World Bank loan documents.

system, there is little alternative. In Korea, the extensive links between banks and enterprise restructuring, and the relatively limited institutional development of these banks, means that the governments will have to play a larger role in enterprise restructuring. Otherwise, banks would be destabilized by large amounts of distressed loans that need to be swapped into equity, and these large equity holdings cannot be disposed of quickly. Furthermore, the broad social and political consequences associated with enterprise restructuring may prevent banks from forcing sufficient enterprise restructuring.

In the case of Korea, a mix of centralized–decentralized approaches could be attractive. The asset management company, KAMCO, or new institutions, could take over some of the distressed assets from financial institutions; once corporations undertake adequate operational restructuring, they could be provided with financial relief through conversion of debt-to-equity claims. A possible model to further accelerate corporate restructuring is the procedure introduced in England to reorganize firms in the 1990s. These so-called London Rules are now being applied in Korea and other East Asian countries. The equities held by these institutions could be the basis for a funded private pension system and a structure which relies more on capital markets, thereby improving corporate regulation. This would reduce debt-equity ratios, and create a new class of outside shareholders who would monitor corporations and balance the role of founders and insiders in the management of chaebols.

The tasks for the medium-term for the financial and corporate sectors are to improve corporate governance of non-financial and financial institutions; develop capital markets; enhance the overall incentive framework for financial institutions; and create more balanced financial systems through the encouragement of capital markets. As it has been for other countries that have gone through a financial crisis, financial sector restructuring in East Asia will be a prolonged process. The main concern is how to prevent the emergence of perverse incentives that would undermine future financial sector development.

East Asia's problems cannot be solved in the banking sector alone. The banks' problems are intricately linked with those of corporations. A solution for financial sectors in distress can only be considered after understanding the problems of the corporate sector. This will be discussed in the following chapter.

Notes

1. See East Asia Miracle, 1994 and Stiglitz and Uy, 1996.

2. Classens and Glaessner, 1997.

3. Total intermediation costs are a function of various factors, including the operational efficiency and competitiveness of the financial sector. Using data as reported in individual annual commercial banks' reports, operating costs (overhead figures) for banks in East Asia do not suggest gross inefficiencies compared to developed countries. Also, compared to developing countries, East Asian banks do not appear to have much higher costs. Reflecting this, it appears that measured profitability over this period was higher in many East Asian banks (with the notable exception of Korea) than in other countries.

4. Claessens and Glaessner, 1998.

5. Kaminsky and Reinhart, 1997 confirm this finding for a wider set of countries.

6. It should be noted that data on real estate lending are not comparable across countries and in several countries likely underestimate the exposure of the banking system to the real estate sector (for example, as loans to developers are not classified as lending for real estate).

7. A private sector committee from Indonesia discussed, with a consortium of creditors a framework which did not include any sovereign guarantees, the treatment of the external debts incurred by its corporations, amounting to about US$70 billion. Agreement on the framework was reached, in principle, on June 4, 1998.

Corporations in Distress

In a dusty outskirts of Surabaya, Indonesia's second largest city, a shrimp-processing plant is humming with activity. Hundreds of young women bend over frozen shrimp, sorting them for shipment to Japan and Europe. They are among the thousands who work for Sekar, among Surabaya's best-known business groups, owned by the five Harry brothers. Until 1994, the Sekar companies followed what was probably the smartest strategy for Indonesian businesses in general: They tapped the country's abundant natural resources to earn valuable foreign exchange. Then the Harrys strayed and began to acquire more glamorous but less-productive assets far from their core business. Sekar's profits amounted to no more than US$13 million in 1996, but Sekar was able to borrow 30 to 40 times that amount from international investors in a few months the following year, expecting to roll this debt when payments came due. The sky fell on Sekar on August 14, 1997, the day Indonesia abandoned its crawling-peg exchange rate system. Sekar, like many Indonesian companies, had bet that the rupiah would maintain its decade-long slow depreciation against the dollar. But the rupiah sank like a stone, taking with it Sekar's fortunes. Now Sekar—like its shrimp sorters—is in slushy water. It is in negotiations with banks and institutional investors to restructure its obligations.—Henny Sender, "Game Over," Far Eastern Economic Review, January 22, 1998.

Cracks in the foundation of corporate performance in East Asia were beginning to appear well before July 1997, but the crisis plunged the corporate sector into a chasm. The financial distress in the corporate sector has become systemic. It is estimated that two-thirds of the firms in Indonesia suffered losses that exceeded their equity; the figure is two-fifths in the Republic of Korea, and one-fourth in Thailand. Interest rates have increased by 5 to 10 percentage points, exchange rates have depreciated by roughly 30 to 40 percent, domestic banks loans are scarce, and foreign capital has dried up.

High leverage, short-term loans, and unhedged foreign borrowing created a set of risks that proved ruinous under the new prices of foreign and domestic capital. As long as growth rates were high, the risks associated with heavy borrowing on a narrow capital base appeared manageable. Many corporations continued to borrow in the face of declining profitability in the mid-1990s. The sudden depreciation and interest rate surge, capped by a fall in demand, converted a weakness into a fatal flaw. High-performing, well-managed companies were suddenly on the verge of collapse along with poorly performing companies.

Today, the most difficult task for policy makers is to provide the right policy framework to untangle the solvent firms from the insolvent, and to stabilize and resurrect viable firms. In the short term, this requires policy measures to restructure the corporate sector, and in the medium term, changes to corporate governance and the link between the corporate and the financial sectors. Strong government leadership has become essential.

The build-up of vulnerabilities in the corporate sector

Behind the miracle. The high leverage ratios that left the corporate sector vulnerable to shocks was the logical outcome of East Asia's growth strategy and system of corporate finance. Over the past decades, East Asian governments followed an aggressive export-oriented strategy. They provided incentives to exporters, such as directed credit, subsidized loans, and tax breaks. Firms required massive resources to continuously upgrade technology and remain competitive in global markets. Retained earnings were insufficient to sustain such an ambitious strategy and equity markets were not well developed; as a result, firms borrowed heavily. A momentum built up—growth was rising and savings increasing—swelling the flow of loans from banks to firms. This financial structure required cooperation among corporations, banks, and the government. This strategy created high leverage ratios, but the govern-

ment provided a cushion against systemic shocks. It also propelled East Asia through a meteoric rise in technology, productivity, and standard of living that surpassed virtually all other countries.

In the 1980s, East Asian governments began to progressively disengage from financial decisions that affected industrial policy. This strategy was beneficial to economies in the process of globalization, but it created a vacuum in institutional and regulatory support. During the 1990s, Asian governments undertook radical financial deregulation: They removed or loosened controls on corporations' foreign borrowing and abandoned coordination of borrowing and investment, but failed to strengthen bank supervision (see box 4.1). After restrictions on foreign borrowings were relaxed, domestic corporations discovered that foreign investors were eager to lend cheaply to East Asian companies. Lenders and borrowers assumed that the fast economic growth would continue and that the exchange rate would remain stable; foreign lenders also ignored their own prudential limits on lending to highly leveraged companies because East Asia was only a small share of their portfolio and they wanted the business. Foreign debt, mostly private and short term, rose and large unhedged positions developed.

Meanwhile, the pattern of investment was undergoing a change. In the 1980s, the investment surge was primarily directed toward tradables, but in the 1990s, most investment shifted toward nontradables, particularly real estate and infrastructure. In an open economy with a fixed exchange rate policy, prices of tradables cannot rise due to import competition. Consequently, especially in Thailand but also in other crisis countries where the nominal exchange rate was fixed and domestic prices were rising, the gap between the price of tradables and nontradables was considerable. This shift in investment increased corporations' and financial institutions' vulnerability to cyclical downturns.

A fragile financial basis for globalizing economies

In the five years preceding the crisis, the East Asian corporate sector's financial performance varied by country. It is clear that while a crisis was building quickly in Thailand, and to some extent in Korea, other countries—Malaysia, Indonesia, and the Philippines—seem

Early warnings signs in Korea

The Korean government's intervention policy, most notably the Heavy and Chemical Industrial (HCI) promotion, has often been evaluated in terms of the success or failure of industrial policy. Another important approach in evaluating intervention policy, however, would be to estimate the financial cost associated with intervention. Government intervention incurred direct costs in the form of subsidies to strategic sectors through policy loans and tax exemptions, especially during the 1973–79 period of HCI promotion. Intervention also incurred indirect costs in the form of accumulated nonperforming loans and the resulting portfolio difficulties of commercial banks.

The promotion of the HCI sector was supported by a broad range of fiscal and financial instruments. Government funds devoted to the HCI sector amounted to 5 percent of the total budget during the promotion period. In addition, profit tax exemptions of up to 100 percent for the first three years and 50 percent for the next two years provided effective tax rates that were about two-thirds lower for strategic industries; at the height of the HCI drive in 1977 about 82 billion won in tax revenues (3 percent of total taxes collected) was lost. The predominant source of financial support for HCI was preferential policy loans directed to key industries. In 1977, for example, 45 percent of the total domestic credit of the banking system was engineered to directly support the HCI sector. Implicit interest subsidies to the HCI sector in 1977 alone were an estimated 75 billion won, or 0.4 percent of gross national product (GNP) (calculated by applying an interest differential of 3 to 4 percent compared with the interest rates

for general bank loans). However, this estimate should be considered the lower bound. Those industries whose credit was squeezed were forced to borrow at curb market rates. The real effective subsidies for HCI borrowers were therefore as high as 3 percent of GNP.

As the Korean government opted to bail out the struggling heavy machinery and shipbuilding industries in the mid-1980s, nonperforming loans of commercial banks accumulated rapidly, and bank profitability seriously deteriorated. During 1986–87, for example, the share of nonperforming loans in total assets reached almost 10 percent. Between 1985 and 1988, 78 corporations were "rationalized." During this process, write-offs of principal alone amounted to US$1 trillion won, or about 1 percent of GNP in 1985. Far more damaging was the further issuance of subsidized funds to rationalized firms and the preferential access to Central Bank discounts, which hurt the balance sheet.

As the Korean government abandoned HCI preferences during the traumatic economic adjustments of 1979–81, the direct costs of interest subsidies all but disappeared. In contrast, it has taken more than a decade to begin to deal with the indirect costs in the form of accumulated nonperforming loans and squeezed bank profitability. Not until 1993 did the Korean government announce a medium-term plan to transfer policy loans, which still account for more than 40 percent of total domestic credit, to separate accounts and to handle the financing of these loans through the budget at as-yet-to-be-determined costs.

Source: World Bank, *The East Asian Miracle*, 1993.

---

to have been affected by the crisis through contagion. Thai corporations' performance deteriorated sharply, particularly since 1995. Korean corporate financial strategy—high leverage, low profitability—was risky but stable until 1995, when the share of firms with inadequate interest coverage increased. By contrast, financial indicators in the corporate sector in Malaysia, Indonesia, and the Philippines were stable in the years preceding the crisis.

Prior to 1997, leverage rose sharply in some crisis-affected countries[1] (see figure 4.1). Between 1991 and 1996, it doubled in Thailand and Malaysia and increased by one-third in Korea. This increase was particularly risky for Korean corporations, given their already high degree of leverage compared with the other East Asian countries. By 1996, leverage for the median firm had reached 340 percent in Thailand and 620 percent in Korea; leverage was lowest in the Philippines, where the magnitude and duration of the lending boom had been smaller than its neighbors (by

contrast, leverage was 80 percent, on average, in the United Kingdom, 100 percent in the United States, and 160 percent in Japan). Meanwhile, profitability was decelerating. From 1991 to 1996, return on assets dipped from 8 to 1 percent in Thailand (see figure 4.2); it declined at a slower pace in Indonesia, the Philippines, and Korea. Only in Malaysia did profitability increase.

The degree of <u>leverage</u> was unhealthy and the declining profitability was disturbing, but the combination was dangerous. Correlations[2] between corporate performance and leverage in all nine East Asian countries show that higher leverage was statistically linked to worse performance (see figure 4.3). This link is strong in Thailand and weaker in the Philippines, but it demonstrates that just before the crisis, corporate and financial sector governance allowed poorly performing firms to get more than their share of financing. This probably made the crisis more severe. However, not all highly leveraged firms were inefficient; some were growing fast, and higher growth was associated with higher earnings (see figure 4.4).

The corporate sector's excessive borrowing compared with its earnings meant that a large share of profit went to cover interest costs (see figure 4.5). Thailand is a good example of this phenomenon. In 1991, its interest coverage resembled that of its regional neighbors, except for Korea; but by 1997, its interest costs had shot up in relation to profits—more than two-thirds of all profits of listed Thai firms went to cover interest expenses. In contrast, the interest coverage of its neighbors remained relatively stable. Since 1995, the share of troubled firms—those whose interest expenses exceeded profits—had risen dramatically for Thailand and Korea (see figure 4.6). By 1997, over one-third of all firms in both countries were in this position. Although this signaled a difficult period, it did not portend a crisis. The situation remained stable in Malaysia and the Philippines, and Indonesia had the lowest share of troubled firms.

Financial performance varied across sectors and mirrored concerns emerging in macroeconomic performance and competitiveness. For example, in Thailand, investment in the construction sector was fueled by real appreciation and necessitated large borrowing. As a result, construction became the most highly leveraged of all sectors—406 percent in 1996 (see figure 4.7). In Korea and Thailand, electronics registered the lowest profitability in 1996 (see figure 4.8), and in Korea, the electronics industry also had the highest share of firms unable to cover interest on loans (see figure 4.9). This reflects Korean firms' "big bets" strategy to compete with the multinationals for market share by developing their own products. This led to massive reductions in Korea's electronics prices in 1996 and reduced profitability in the sector (see chapter 2). In contrast, Malaysia and the Philippines enjoyed high profitability in the electronics sector in 1996 because they were part of a more diverse production network through their direct links to multinationals.

Weak corporate governance: Fertile ground for high leverage

The East Asian crisis has underlined the importance of the rules, norms, and organizations that govern corporate behavior and define accountability to investors.

East Asian corporate finance markets typically are dominated by banks. Because securities markets require a more sophisticated institutional and regulatory framework, bank dominance of corporate finance is probably the best way for developing countries to grow, provided that they are not subject to undue state influence, are exposed to competition, and are prudently regulated. East Asian countries did not meet these criteria and failed to adjust as globalization was taking place (see box 4.2).

Weak incentives to improve governance. The financial sector has functioned poorly in East Asia. In a market that had ample liquid assets, lack of market discipline plus comfortable relations between bankers and their corporate clients led to increased loans to firms with high leverage and low profitability. In this environment, there was little to gain from improving disclosure and corporate governance. The lack of market discipline resulted in part from policy residues and from optimism about future growth, as underlined by the five areas below.

First, unhealthy links between government and banks and between banks and clients are left over from early policies. In East Asian countries, there was political pressure to make loans to favored firms, and implicit state guarantees to back bank loans to priority sectors, to ensure bank depositors were not at risk. These measures weakened bank incentives to ensure that their loans were creditworthy. Banks continue to suffer from heavy state involvement. For example, in Korea, the state took control of the banking system in the early 1960s. It has privatized and sold off its major shareholdings in most banks over the past 20 years, but its influence over banks' operations has remained substantial. Moreover, there may have been interlocking ownership and other interrelationships between banks and corporations, which reduced market discipline.

Second, domestic institutional investors are scarce and underdeveloped. Institutional investors help establish private market incentives to adopt good corporate governance practices, such as managing pension funds. In Thailand, the mutual fund industry represented only 7 percent of total Stock Exchange of Thailand (SET) trading in 1996. Pension funds are weak in most East Asian countries. Until recently, these funds have been constrained by the lack of formal institutional pension arrangements and by restrictive asset allocation regula-

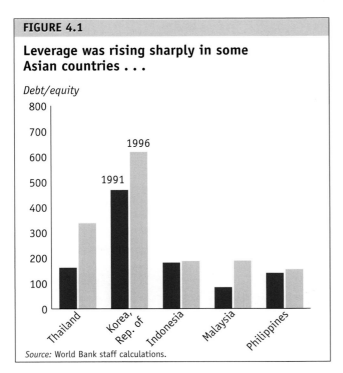

FIGURE 4.1

Leverage was rising sharply in some Asian countries . . .

Debt/equity

Source: World Bank staff calculations.

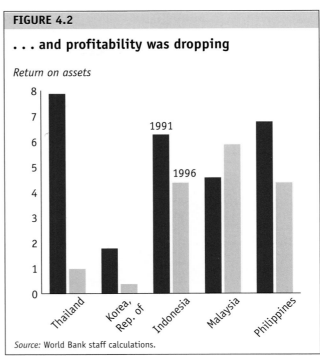

FIGURE 4.2

. . . and profitability was dropping

Return on assets

Source: World Bank staff calculations.

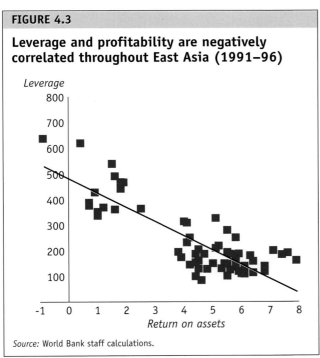

FIGURE 4.3

Leverage and profitability are negatively correlated throughout East Asia (1991–96)

Leverage

Return on assets

Source: World Bank staff calculations.

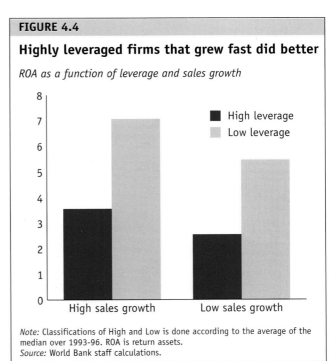

FIGURE 4.4

Highly leveraged firms that grew fast did better

ROA as a function of leverage and sales growth

■ High leverage
▨ Low leverage

High sales growth   Low sales growth

Note: Classifications of High and Low is done according to the average of the median over 1993-96. ROA is return assets.
Source: World Bank staff calculations.

tions. Provident funds for employees in both government and public companies have only been established recently and are still largely restricted to government paper and cash.

Third, the role of foreign banks in funding East Asian firms, and in imposing international standards of gov-

ernance, is still limited. Asian countries stand out from other emerging markets because of the relatively low share of the market held by foreign banks. This is partly because foreign bank entry is restricted; most East Asian countries have a policy of protecting domestic banks from foreign competition by severely limiting the

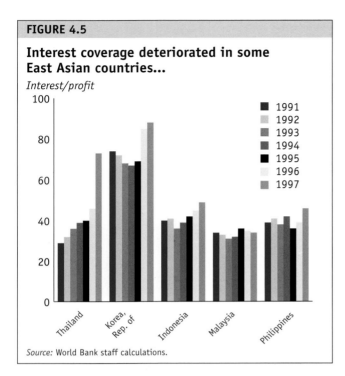

FIGURE 4.5

Interest coverage deteriorated in some East Asian countries...

Interest/profit

Legend: 1991, 1992, 1993, 1994, 1995, 1996, 1997

Source: World Bank staff calculations.

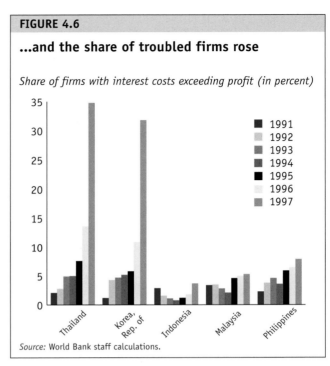

FIGURE 4.6

...and the share of troubled firms rose

Share of firms with interest costs exceeding profit (in percent)

Legend: 1991, 1992, 1993, 1994, 1995, 1996, 1997

Source: World Bank staff calculations.

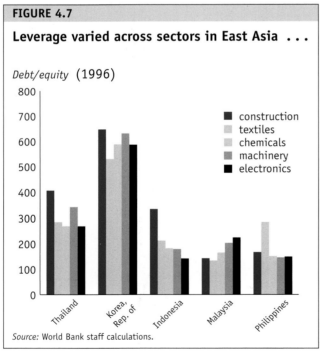

FIGURE 4.7

Leverage varied across sectors in East Asia . . .

Debt/equity (1996)

Legend: construction, textiles, chemicals, machinery, electronics

Source: World Bank staff calculations.

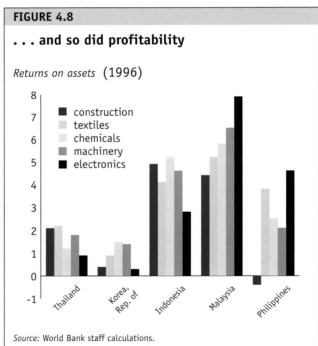

FIGURE 4.8

. . . and so did profitability

Returns on assets (1996)

Legend: construction, textiles, chemicals, machinery, electronics

Source: World Bank staff calculations.

number of new banking licenses granted to foreign banks. In addition, foreign banks that do have a presence in East Asian countries are often content to lend only to the best firms and for the safest purposes, such as trade finance. In some countries like Indonesia, a large share of foreign funding was off shore and did not

have the same monitoring impact as a foreign bank located within the country.

Fourth, foreign institutional investors have not been insistent on corporate disclosure and governance. Foreign investors may have believed, as many domestic participants in East Asian financial markets apparently did, that states had a "too-big-to-fail" policy for all

BOX 4.2

Corporate governance in East Asia and other emerging economies (prior to July 1997)

Variables	Description/Effect	Indonesia	Malaysia	Philip-pines	Thailand	Mexico	India	Pakistan
Right to Call Emergency Shareholder Meeting (percent share capital to call meeting)	Facilitates shareholder control.	YES 10	YES 10	YES 10	YES 20	YES 33	YES 10	YES 10
Right to Make Proposals at Shareholder Meetings.	Facilitates shareholders control; increased opportunity to prevent biased decisions by insiders.	YES	YES		YES	NA	NA	
Mandatory Shareholder Approval of Interested Transactions.	Protects against abuse and squandering of company assets by insiders.	YES	YES	YES	YES	NA	NA	
Preemptive Rights on New Stock Issues	Protects against dilution of minority share-holders; prevents insiders altering ownership structure.	YES		YES	YES	NA	NA	
Proxy Voting	Facilitates shareholders control.	NO	YES	YES	YES	NO	YES	YES
Alternative Dispute Resolution mechanism	Facilitates shareholder control.	YES						
Mandatory Reporting by Large Shareholders	Disclosure of transactions by large shareholders protects against abuse by insiders.	YES	YES	YES	YES			YES
Ownership concentration in the 10 largest private firms (percentage owned by the 3 largest shareholders)		62	NA			64	43	49
Penalties for Insider Trading	Protects against use of undisclosed information at the expense of current and potential shareholders.	YES	YES	YES	YES			
Provisions on Takeovers Legislation	Protects against violation of minority share-holders' rights.		YES	YES	YES			
Mandatory Disclosure of Non-Financial Information	Both financial and non-financial information data are important to assess a company's prospects.	YES	YES		YES			
Mandatory Disclosure of Connected Interests	To protect against abuse by insiders.			YES	YES			
Mandatory Independent Board Committees	If composed of independent directors, audit and remu-neration committees protect against insider abuse.		YES		YES			
Mandatory Shareholder Approval of Major Transactions	Protects against abuse by insiders. Protection can be enhanced through supra-majority voting.	YES	YES	YES	YES	YES		YES
One Share–One Vote	Basic right; some shareholders may waive their voting rights for other benefits such as higher dividends.	NO	YES	NO	NO	YES	NO	NO
Allow Proxy by Mail	Facilitates shareholder control.	NO	NO	NO	NO	NO	NO	NO
Shares Not Blocked Before Shareholder Meeting	Mandatory depositing of shares prior to share-holder meeting makes shareholder control more difficult.	YES	YES	YES	YES	YES	YES	YES
Cumulative Voting for Directors Allowed	If shareholders can cast all of their votes for one candidate, it increases probability of out-side directors.	NO	NO	YES	YES	YES	NO	NO
Automatic Stay on Assets if filed for bankruptcy	Shields shareholders from creditors	NO	NO	YES	NO	NO	YES	NO
Oppressed Minorities Mechanisms	Shareholders' right to a judicial venue to chal-lenge insider decisions or to request company purchase their shares if they object to funda-mental changes.	NO	YES	YES	YES	YES	NO	NO

large domestic firms, and thus were more confident about the threat of losing their money than they would have been otherwise. Moreover, the fast real growth rates and rapid upward movement of stock prices in East Asian countries may have created a euphoria among international investors that caused them to overlook or disregard the deficiencies in governance practice in these countries.

Fifth, market and regulatory institutions that play an important role in industrial countries in facilitating and creating incentives for market discipline are not yet fully developed. For example, Thailand's single credit-rating agency TRIS was established in the 1990s and is still considered by the market to be developing expertise. The nascent regulatory framework further aggravated this lack of market institutions. A modern Thai legislative regulatory framework was promulgated in 1992, the same time that the Securities and Exchange Commission (SEC) was established. By 1997, Thailand built the legal and regulatory basis for modern capital markets, but this process has been gradual. During the transition period, capital markets did not adequately perform their signaling and monitoring functions.[3]

Ownership structure also led to high leverage. The most common organizational form in East Asian corporations is the diversified conglomerate that is closely held, controlled, and managed by a family. In Korean "chaebols," as these diversified conglomerates are known, families own far less than 50 percent of chaebol-related companies, but they have almost total control over their business groups. Interlocking ownership allows them to control related companies with very little equity of their own: with each member company holding every other company's shares, the percent ownership of a chaebol family (in relation to total outstanding capital) increases. Although the founder of the company and their immediate relatives may hold a small percentage of outstanding shares (between 3 and 15 percent) of the chaebol business groups, the intercorporate ownership of chaebol member companies increases the total inside ownership to 30 to 60 percent (see table 4.1).

Ownership concentration has benefits and costs. On the benefits side, it has been associated with firms enhancing their efficiency of operations and investment. On the costs side, it may lead controlling owners to expropriate other investors and stakeholders and

pursue personal nonprofit maximizing objectives, and it may impede the development of professional managers, who are required as economies and firms mature and become more complex. Empirical evidence shows an inverted "U"-shaped relationship between the degree of ownership concentration and profitability.[4]

TABLE 4.1
Ownership of Korean business groups by insiders
(percent of common shares held)

Business group	Founder	Relatives	Member companies	Total
Hyundai	3.7	12.1	44.6	60.4
Samsung	1.5	1.3	46.3	49.3
LG	0.1	5.6	33.0	39.7
Daewoo	3.9	2.8	34.6	41.4
Sunkyong	10.9	6.5	33.5	51.2
Sangyong	2.9	1.3	28.9	33.1
Hanjin	7.5	12.6	18.2	40.3
Kia	17.1	0.4	4.2	21.9

Source: Lae H. Chung, Hak Chong Lee, and Ku Hyun Jung, 1996.

Correlation of leverage on initial ownership concentrations among Thai corporations for 1992 and 1996 shows a perverse effect in both years (see table 4.2): Firms with more concentrated ownership have higher leverage, even when adjusted for cross-sector and size differences. This effect almost doubled in magnitude between 1992 and 1996, suggesting that corporations with high initial ownership concentration crowded out

FIGURE 4.9

The share of troubled firms varied across sectors in East Asia

Share of firms with interest costs exceeding profit (1996)

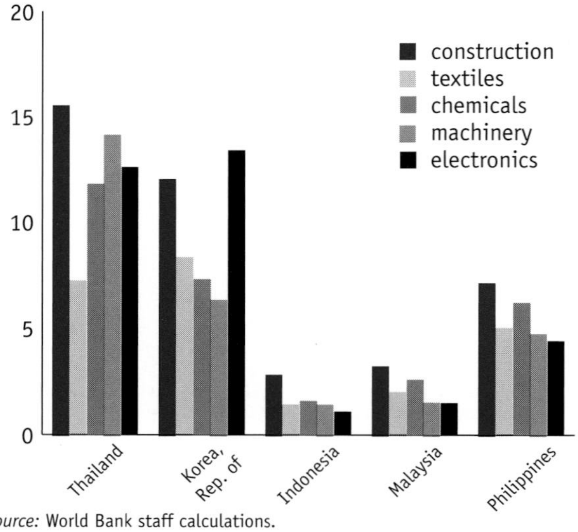

Source: World Bank staff calculations.

the less-connected firms. Correlation of ownership concentration and profitability were positive in 1992 but turned negative in 1996 (although these numbers were not significant). The costs of concentrated ownership in East Asia were further exacerbated by the relationship between corporations and banks, particularly corporations owned by families that had controlling interests in banks. This relationship increased the possibility of easy borrowing, thus leading to higher leverage without a direct link to performance and to a slower response to changing market conditions, which increased the risk of expropriation.

TABLE 4.2

Ownership concentration has a perverse relationship with profitability and leverage in Thailand (correlation coefficients)

	Profit 1992	Profit 1996	Leverage 1992	Leverage 1996
Ownership concentration, 1992	0.202*	-0.063	0.151*	0.287*
	(0.084)	(0.090)	(0.044)	(0.072)
R-squared	0.176	0.163	0.265	0.322

Note: (1) Profitability is EBIT (earnings before interest and taxes) over sales; ownership concentration is the share of top five owners, leverage is debt over equity; (2) sector dummies included for profit correlations, sector and size dummies included for leverage correlations; (3) number of observations is 236; (4) standard errors in parentheses; (5) * denotes significance at 5 percent level.
Source: Alba, Claessens, and Djankov, 1998.

After the crisis: Assessing the damage

After the crisis, bankruptcies spread throughout the region. Indonesia was hardest hit, followed by Korea; Malaysia and the Philippines suffered less, and Thailand was in the middle. Several shocks affected East Asian firms' ability to service their debts and access new financing. First, international investors lost confidence and retrieved their capital; from a net inflow of US$97 billion in 1996, foreign capital dropped to a net outflow of US$12 billion in 1997. Second, unprecedented exchange rate devaluations of about 40 percent aggravated the debt service burden and created extensive losses. Corporate foreign debt was largely unhedged because firms believed that the exchange rate would remain stable, as it had for the past decade (see box 4.3). Third, tight monetary policies, enacted to preempt further devaluation and avoid hyperinflation, created an unprecedented 5 to 10 percent hike in interest rates. This was particularly damaging to the highly leveraged corporate sector.

How many firms will go out of business in the crisis countries? As a result of the interest rate and exchange rate shock, corporations throughout East Asia now face a substantially higher leverage and share of foreign debt. A simulation of the impact of these shocks shows that leverage is 50 percent higher and the share of foreign debt is 30 percent larger without changing the financial structure of the corporations.[5] On average, firms lost about half of their equity as a result of these shocks, and for almost one-third of them, losses exceeded their equity. Their return on assets dropped from 3 percent to -4 percent (see figure 4.10), and two-thirds of them now have negative returns on assets— four times more than before the crisis.

The extent of the damage has reached systemic proportions in most crisis countries. Indonesia suffered the most, with nearly two-thirds percent of its corporations declaring bankruptcy (see figure 4.11). This is due more to the magnitude of the exchange rate shock than to the financial performance of the corporate sector prior to the crisis. Korea came second. Prior to the crisis, Korean corporations were highly exposed to risks because of their high leverage and low profitability. Malaysia has suffered much less, as its initial conditions were better. Thailand and the Philippines were in-between.

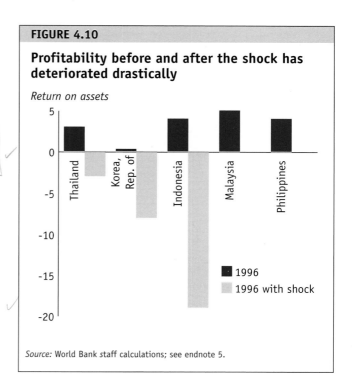

FIGURE 4.10

Profitability before and after the shock has deteriorated drastically

Return on assets

- 1996
- 1996 with shock

Source: World Bank staff calculations; see endnote 5.

The immediate agenda: Restructuring banks and corporate systems

The breadth and depth of corporate distress in East Asia is unprecedented in recent economic history. The level and structure of corporate debt, the number of debtors and creditors involved, and the weak legal environment make corporate restructuring a daunting challenge. At the same time, financial resources are scarce: The banking system is distressed and stock markets are depressed, thus offering little hope for a quick resumption of financial flows to enterprises. Solving this systemic crisis requires a comprehensive strategy in which enterprises, financial institutions, and governments work together.

Financial restructuring determines not only the allocation of current losses but also the distribution of ownership and future control in the economy. In seeking a solution, the involved parties have to address a range of questions: Who will bear the costs of restructuring? How fast can a solution be implemented? How can the survival chances for viable enterprises be maximized? The answers to these questions determine the extent of government involvement in the restructuring process. Whatever the approach, the government has to ensure that any solution contributes toward improved enterprise governance, thereby reducing the chances of repeated financial sector bailouts. Each country in East Asia is going through a process of political and economic choices toward these outcomes.

History is a poor guide to finding a solution, but it offers useful lessons. The experience in Latin American countries in the 1980s (see box 4.4) and the transition economies in the 1990s shows that managing a crisis of the magnitude that the five East Asian countries now face is fraught with problems, often is expensive to the taxpayer, and often leads to repeated problems after only a few years. The experience of the Nordic countries provides another example. These countries resolved their crises relatively quickly after their financial system weakened in the early 1990s. Their problems were confined to only part of the banking system,

---

BOX 4.3

Which firms borrow in foreign currency?

A 1997 survey of 840 firms in Thailand showed that less than 4 percent of firms with foreign currency debt were fully hedged and only 17 percent were partially hedged. However, the results of the survey suggested that the problem of unhedged borrowing was not as extensive as might be feared. The large majority of firms borrowed only in baht, and those that did borrow in foreign currency were generally the more efficient ones (they are predominantly large exporting firms with ties to foreign companies, and they have better adjusted to the crisis with higher capacity utilization and cuts in employment). The issue is more severe in the private banking sector than in the manufacturing sector, particularly with financial institutions that borrowed from foreign investors and then loaned this money to domestic companies and assumed the exchange risk themselves.

Profile of Thai firms that borrow in foreign currency

	Firms that borrow	Firms that do not borrow
Frequency	25%	75%
Financial indicators		
Short-term debt/total debt	77%	85%
Long-term debt/total assets	15%	10%
Debt-equity ratio	312%	236%
Firm characteristics		
Number of employees	818	139
Share that export	88%	46%
Share that are joint ventures	60%	6%
Response to the crisis		
Current capacity utilization	70%	61%
Share with fewer workers	48%	57%
Optimistic for future growth	37%	19%

Source: Dollar and Hallward-Dreimeier, 1998.

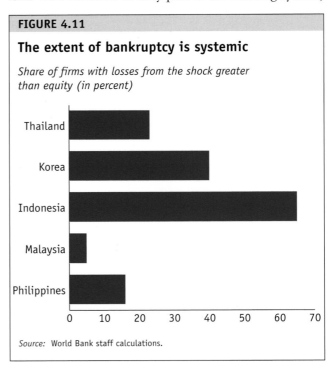

FIGURE 4.11

The extent of bankruptcy is systemic

Share of firms with losses from the shock greater than equity (in percent)

Source: World Bank staff calculations.

BOX 4.4

Corporate restructuring in Chile in the 1980s

In 1973, about 600 firms, including all those in the financial sector, were managed by the state in Chile. At that time, the state-owned enterprises (SOEs) had an average debt-to-asset ratio of about 40 percent, which was not considered excessive and did not differ from that of the private sector. During 1974, the year in which the massive Chilean privatization program began, negative real interest rates of over 38 percent contributed to reduce the debt-to-asset ratio even further. Thereafter, a government policy to severely restrict all new SOE investments had the same effect.

The roots of the debt problem. The debt-led privatization of about 550 SOEs during the 1970s generated a large number of highly indebted corporations and holding companies, because the purchasers of SOEs did not have any significant financial assets. In fact, the average debt-to-asset ratio of a sample of privatized corporations increased from 40 percent in the early 1970s to about 70 percent during the early 1980s. Furthermore, about 40 percent of the debt was expressed in U.S. dollars, because Chile was experiencing high capital inflows at relatively low interest rates. An economic and financial crisis broke out in Chile in 1982–83. The government devalued the currency, while real interest rates on loans, which had come down to about 10 percent, increased during 1981–82 to over 35 percent. The devaluation and the extremely high real interest rates increased indebtedness ratios even further. This deeply affected the debt-service capacity of a large proportion of Chilean firms and therefore the solvency of the financial system. The government estimated that the losses to the financial system amounted to between US$2.5 to US$4.0 billion dollars, far exceeding its capital of about US$1.1 billion. Under such circumstances, the government decided to intervene in the operations of a large number of financial institutions in early 1983, including the largest commercial banks. In so doing, the government took over again the management of the key financial institutions of the country and, indirectly, that of a number of formerly privatized SOEs.

The solution. To solve the problem, a number of measures were applied. First, the intervention by the government into a large portion of the financial institutions with estimated negative net worth reduced depositor risks and loan roll-overs, and therefore reduced the high real interest rates and the debt burdens of corporations. In addition, a few financial institutions—those with the lowest relative net worth—were closed down, limiting depositor losses to 30 percent, and the remaining losses were absorbed by government. Second, the Central Bank swapped the bad loans of the remaining financial institutions for interest-bearing long-term bonds; in this way, the state cleaned up the banks' balance sheets. The financial institutions, however, assumed the responsibility of repurchasing the bad loans, committing themselves to repurchase on an annual basis bad debts amounting to a given proportion of profits (in most cases, over 70 percent). The Central Bank also advanced resources to the financial institutions to allow them to refinance loans to all medium- and small-sized producers of goods and services and mortgage lenders. This program's objective was to improve the quality of these loans as well as to clean the balance sheets of most of these producers. Finally, having improved the quality of the loan portfolio, the financial institutions had to be adequately capitalized. This was done by issuing new bank shares to groups of new shareholders who would assume a controlling interest. However, in the case of the two largest commercial banks the government used "popular capitalism" to recapitalize them: The banks issued shares that could be paid over 15–20 years at zero real interest rate. Furthermore, the value of the purchase could be deducted from the income tax base each year until repayment of the loan, as long as the purchaser held onto the shares.

Source: Hachette, D. and Luders, R. (1992); Luders, R. (1998).

whereas the corporate sector remained, with some exceptions, largely viable. Nevertheless, it cost Sweden 5 percent of GDP and Finland 10 percent of its GDP to support their banks.[6]

Options for restructuring

In general, solutions have worked best if the costs to taxpayers were minimized, shareholders were hit hardest, banks did most of the enterprise restructuring, and the government did not end up as the owner of a large number of banks and enterprises. There are three approaches that governments can consider: the market-based approach, the recapitalized bank-led approach, and the government-led approach.

The market-based approach. The market-based approach aims to use mainly market forces to restore enterprise profitability and bank capital. Measures include operational restructuring of enterprises, leading to higher efficiency and profitability; foreign inflows for new investments; asset sales to foreign and domestic investors; offshore and domestic equity issuance; and debt restructuring. Most of these approaches are currently being used on a case-by-case basis in East Asian countries. Governments have also undertaken several steps to enhance the enabling environment, including allowing hostile mergers and acquisitions and liberalizing rules pertaining to foreign investment. Furthermore, East Asian governments have promoted capital market development, such as through the adoption of a mutual funds law in Korea.

The market-based solution limits the burden on taxpayers, reduces the likelihood that governments end up as the primary owner of banks and enterprises, and helps deepen capital markets. But for many East Asian countries, the market-based solution is unlikely to reduce debt to sustainable levels for many companies in the next few years. In Korea, for example, a purely market-based solution will result in an average debt-equity ratio of the largest chaebols of over 400 percent in the year 2000. Although Korean firms traditionally have had high debt-equity ratios, this 400 percent ratio is above historical levels and much higher than is common in market economies. Similar analysis applies to other East Asian countries.

This high debt burden leaves enterprises vulnerable to fluctuations in the interest rates and market conditions. Also, the market-based solution cannot fully resolve non-performing loans of the banking systems in the next few years. This could aggravate the lack of confidence among investors and the credit crunch felt by firms. The market-based solution also is unlikely to lead toward a more balanced corporate financing structure, which relies more on capital markets and improved corporate governance. Policies involving accelerated restructuring tackling stocks, that is, restructuring of liabilities, are probably necessary in most East Asian countries.

The recapitalized bank-led approach. The recapitalized bank-led approach recapitalizes banks that then take the lead in corporate restructuring, including financial restructuring. Under this approach, the government recapitalizes banks based on an ex-ante assessment of their losses. Individual banks or groups of banks then work out the problem debts and take charge of the operational restructuring, possibly providing working capital during restructuring. The government does not intervene directly in corporate sector restructuring. Most transition economies used this approach, with the most notable success in Poland.

The recapitalized bank-led approach has several benefits. It can be relatively fast and can signal to the market that the problems are being resolved (dissatisfaction with the slowness of the current approach has been an important concern among foreign investors). As several East Asian governments have already guaranteed the liabilities of banks and other financial institutions, recapitalization by the government would only formalize this process without additional costs. Provided recapitalization is accompanied by substantive changes in the corporate governance and operations of banks, it can be an up-front investment that may ultimately lead to lower costs.

Yet, the recapitalized bank-led approach has risks. Experience shows that governments routinely inject capital into insolvent institutions without creating sufficient change in the bank's governance and operations. To date, most bank recapitalization programs instituted by governments have been generally unsuccessful. When facing the trade-off between maintaining confidence and preserving incentives for good banking (that is, minimizing moral hazard), most countries have favored maintaining confidence by extending large-scale guarantees or continuing to recapitalize banks. In East Asia, the previous role of the government in the real and financial sectors and the large problems in corporate sectors give little guarantee that recapitalization will be either sufficient or occur only once.

The recapitalized bank-led approach has additional disadvantages for the East Asian crisis countries. First, the very high debt-equity ratios require substantial debt-equity swaps. But if banks end up holding large amounts of equity, they can become more vulnerable to stock market fluctuations. Second, many East Asian banks also lack the technical capacity and skills to restructure a large number of enterprises. Although technical assistance is rapidly improving the banks' restructuring capacity, it will take time before they can effectively restructure enterprises. Third, banks may be too weak vis-à-vis corporations in restructuring negotiations. Some enterprises, such as the chaebols in Korea, may be too big to fail because restructuring would have many social and political consequences. This may result in weak corporate restructuring plans and ultimately larger fiscal costs.

The government-led approach. Under a government-led approach, the government or a government agency takes over a large share of distressed assets from the banks and replaces these assets with government bonds or other safe assets, thereby recapitalizing the banking system. The government then tries to restructure the claims and to force corporate restructuring. The main advantage of a government-led solution is that it can be fast and create clarity as it segregates bad bank loans into a new agency. It can also shift the balance of power to creditors in the case of large corporations.

The government-led approach also has risks. First, the transfer of loans breaks the links between banks and corporations, links that may have positive values because of banks' privileged access to information. Thus, large transfers without bank involvement in the restructuring process may create asset value losses. Second, a nonbank entity like an asset management company (AMC) may not have the capacity for the working capital lending that is often required during debt restructuring. Third, and most important, a government-owned agency may have poor incentives to restructure corporations. Even though private managers could run AMCs with various incentive programs built in, the risk of poor management remains. Past experiences, particularly in countries with weak institutions, suggest that many times an agency "sits" on its loans, often in fear of antagonizing the "powers that be," the same powers who often contributed to the bad loans in the past.

Progress in corporate debt restructuring among East Asian countries

Most East Asian countries have yet to complete the formulation of a comprehensive framework of corporate restructuring. Such framework depends not only on the magnitude but also on the characteristics of corporate indebtedness--debt to local banks and debt to external creditors, mostly foreign banks. The relative weight of these debt components varies considerably across countries. In Indonesia, the bulk of external borrowing by the private sector was undertaken directly by the nonfinancial corporate sector. In Korea, where corporate regulations hampered such borrowing, a much larger proportion of external debt was taken on by local banks for lending on to local corporations. In Thailand, both banks and corporations were significant external borrowers, while in Malaysia, external debt is relatively low.

Compared to banking sector reforms, corporate restructuring is largely at the beginning stage. Thailand, Korea, Indonesia, and Malaysia have adopted more aggressive frameworks for corporate debt workouts going beyond market-based, establishing corporate debt restructuring (or arbitration) committees based on the "London rules" so as to give creditors and debtors sufficient incentives to implement voluntary workout. However, the actual process of corporate workouts has yet to begin. Elements of the bank-led and government-led restructuring approach have also been established. In Korea, a fraction of the banks' domestic nonperforming loans have been transferred to the government-owned Korean Asset Management Company (KAMCO), which will restructure them. A bridge bank has further assumed distressed assets of some merchant banks. In Thailand, the government-owned Financial Restructuring Authority (FRA), in competition with the private sector, has acquired assets from defunct finance companies and has been selling some of these assets at auctions. An AMC has also been established to purchase impaired assets. In Malaysia and Indonesia, an AMC also has been set up. To date, the AMCs have been largely inactive in forcing restructuring throughout East Asia.

Progress in dealing with external debt restructuring has been achieved in several East Asian countries. Korea was the first crisis country to restructure its external debt in addressing private sector debt issues. Korean banks converted US$24 billion in short-term nontrade debt of commercial banks into new loans with maturities between one to three years guaranteed by the Korean government. The chief motivation for public intervention was to avoid severe disruptions that failures could cause, because commercial banks form a large portion of the domestic payment system. In addition, the government was willing to intervene because there were only a limited number of creditors and few debtors. Some of the key features of the Korean restructuring plan are (a) the new loans bear interest of 225, 250, and 275 basis points over a six-month London interbank offered rate (LIBOR); (b) each bank is allowed to swap up to 20 percent of its eligible loans for the new one-year loans, and the new two- and three-year loans will carry call options permitting penalty-free repayments at six-month intervals; and (c) the new loans are in the form of transferable certificates.

The Indonesian external debt restructuring agreement resembles Korea's approach, in which bank debt would be rolled over with a government guarantee.[7]

The agreement comprises three components: a framework for restructuring the external debt of corporates, a scheme to repay interbank debts, and an arrangement to maintain trade finance facilities. In addition, it entails the introduction of the Indonesian Debt Restructuring Agency (INDRA) to implement a voluntary program for the provision of foreign exchange availability to Indonesian corporate debtors and their foreign creditors. The government guarantee of foreign exchange risk will form the core framework for corporate debt restructuring. Participation in the agreement is voluntary and decided on jointly by debtors and creditors. INDRA will provide a real exchange rate guarantee and assurance that foreign exchange will be available to service debts without assuming commercial risk. Each debtor and creditor will be expected to renegotiate the debt, agree on possible debt reductions, debt-equity swaps or other debt-reduction techniques, and lower the repayment stream within this framework. To participate, the terms of renegotiated loans must have minimum maturity of eight years with a three-year grace period. The agreement also could include other concessions from lenders on existing debt and efforts by the borrower to repay the debt, including through the sale of assets.

In contrast, the Thai authorities preferred to encourage debtors and creditors to negotiate obligations through voluntary arrangements. The external debt structure of the Thai financial sector is skewed mainly toward commercial bank loans along with loans booked through the Bangkok International Banking Facility (BIBF). A large portion of the BIBF loans were on-lent by domestic banks to the finance companies. Efforts to resolve problems included closing 56 finance companies plus initiatives by the government to restructure the banking system. The restructuring process to date differs by type of financial institution: for insolvent finance companies, closure and disposal of their assets; and for severely distressed banks, nationalization and eventual privatization. For the corporate sector, the government has stated that no public support will be rendered to the bulk of obligations. However, there are several regulatory impediments to voluntary restructuring in Thailand, which are currently being addressed (see box 4.5).

A possible approach for East Asia: The blend variant

Current approaches for solving the economic crisis in most East Asian countries straddle the market-led, bank-led, and government-led approaches. Banks are partially recapitalized and left to work out the distressed assets, particularly the medium-sized and small loans, whereas some larger loans are transferred to AMCs and other agencies, with the AMC leading the restructuring. This approach may work in countries in which the problems are predominantly with smaller corporations which banks can be expected to restructure, or the problems are with a few number of large enterprises, which are difficult to restructure and require government involvement. However, for most East Asian countries, none of these conditions may apply. It is also not clear that these approaches alone will achieve a necessary change in corporate control. In particular, the shortage of equity capital means that in the financial restructuring of enterprises, existing shareholders cannot be completely eliminated, otherwise banks or the government will own much of the corporate sector. Some involvement of current owners may actually help solve the crisis, because they have proprietary information. But leaving current shareholders in control may perpetuate the weak governance that led to the problems in the first place.

Coordination among the government, banks, and corporations. Corporations in the East Asian crisis countries are institutionally stronger than banks and have shown better management. They likely face better incentives than banks in restructuring their businesses. Therefore, they could play an important role. One approach for large, difficult restructuring cases would be building on the strengths of banks, corporations, and governments in an arbitration process. Corporations could be forced to apply to this process by the government limiting any new lending to an enterprise with debt-servicing problems. In the process, an enterprise would have its obligations restructured, providing it undertakes the necessary operational restructuring.

The arbitration process would lead to an operational restructuring plan that would be presented to a committee consisting of creditor banks, the government as represented through an AMC, and other independent members (with various consultants and staff to support such a committee). The plan should indicate the savings achieved through operational restructuring, the selling of assets, the contribution in new equity by the existing shareholders, and the degree and modality of financial relief sought. If the plan is not acceptable, the corporation would be subject to a standard bankruptcy process.

The form of financial relief provided by creditor banks could include extending maturities and interest relief but would contain only limited debt-equity swaps. Banks may provide working capital but under strict conditions (for example, only highly collateralized capital). The AMC or other funds would be allowed to take over some loans from creditor banks and swap some of these claims into equity. By substituting equity for debt, the government would provide the corporation time to overcome its financial problems.

The AMC as a shareholder, perhaps the largest single one, would have the responsibility to ensure that the corporations are properly managed. It is not desirable, however, for the government to become a permanent owner of enterprises. Disposal of the securities acquired—through sales to strategic or portfolio investors—as quickly as possible would be desirable, but this will be difficult given the limited supply of equity capital. There are several options. Equity could be transferred to taxpayers to compensate for the taxes necessary to finance the bank and corporate restructuring, or the shares could be transferred to an institution that promotes public benefits. For example, the shares could become the basis of privately managed, funded pension programs. Other programs would include citizens being encouraged to purchase shares through advantageous financing plans.

There are several potential advantages of the blend approach for the combined problems of corporate and bank restructuring (see table 4.3). First, it would deal with the stock problem within the framework and context of financial sector restructuring, where the governments have already committed resources. Specifically, it would set up improved mechanisms for loss-taking which are insufficient even under London-based approaches. Second, it would not break the link between banks and enterprises, as banks remain significant lenders to viable firms, which may have a positive value, given banks' privileged access to information. Third, it simultaneously includes all parties involved (banks, governments, AMCs, and corporations) and incentives are balanced with an emphasis on corporations to create credible operational restructuring plans. Fourth, by undertaking debt reduction and debt-to-equity swaps by the AMC or other funds and debt restructuring and new money by the banks, the power of creditors over the corporations is enhanced. Finally, if debt-equity swaps are linked to the creation of equity funds (pension funds, mutual funds, etc.), and if foreign investment is encouraged, the creation of a large set of active, outside owners will strengthen corporate governance.

Improving corporate governance

The main lesson from the East Asia crisis is that it is important to take an integrated approach to the issues of corporate governance and financing. The poor system of corporate governance has contributed to the present financial crisis by shielding the banks, financial companies, and corporations from market discipline. Rather than ensuring internal oversight and allowing external monitoring, corporate governance has been characterized by ineffective boards of directors, weak

TABLE 4.3
The impact of restructuring options

	Speed	Limiting fiscal costs	Incentive for enterprise restructuring	Long-term effect on corporate governance	Demand on legal and regulatory environments
Market solution	●	●●●●	●[1]	●●	●
Recapitalized solution	●●	●●●	●●●●[2]	●●●	●●●●
Government-led solution	●●●●	●	●●/●●●	●	●●●
Blend	●●●	●●	●●/●●●	●●●●	●●

[1] Assuming close ties between banks and enterprises.
[2] Provided sufficient incentives for banks' performance.
Note: More dots should be interpreted as "better."
Source: Staff evaluation.

Removing regulatory impediments to voluntary restructuring: Lessons for Thailand and other crisis countries

The most important regulatory policies are to expand the role of foreign investment in the corporate sector; to review tax rules that may discourage debt restructuring, debt-equity swaps, mergers, and acquisitions; and to review bankruptcy legislation so as not to discourage new money to firms in financial distress. Regulatory reforms needed to promote voluntary reorganizations of illiquid but viable corporations include the following steps:

Eliminate tax disincentives to equity restructuring (that is, mergers and acquisitions). An asset transfer or share acquisition in the course of corporate reorganization should not be treated as a taxable event if the corporate seller (or its shareholders) receives not cash but only the newly issued shares or the existing shares of the acquirer. Similarly, a merger should not be a taxable event. Taxes would be paid only on sale of the acquired shares. In addition to relaxing the tax treatment of noncash corporate reorganizations, it would be useful for governments to consider allowing the transfer of tax-loss carry-forwards to an acquirer or postmerger entity.

Temporarily reduce or eliminate tax disincentives to debt restructuring. Currently, a creditor could not discharge debt without creating taxable income (that is, the debt forgiven) for the debtor in Thailand. This reduces the incentive to discharge debt. It is not clear how current tax laws would treat other techniques of debt restructuring (for example, term extensions or rate reductions). Because debt restructuring is likely to be a crucial element of corporate restructuring, governments should identify possible tax disincentives and remove them, at least temporarily. Governments also need to develop criteria for providing tax relief for debt restructuring to discourage tax evasion and avoid providing tax relief for corporations that do not need such relief or that are candidates for court-supervised reorganization or bankruptcy (that is, liquidation).

Eliminate interest deductibility on excessive debt. Because current tax laws may encourage corporations to borrow too much, it would be useful for governments to develop a "thin capitalization" rule. Such a rule would prohibit interest expense deductions on debt above a specified threshold (for example, a debt-equity ratio of 200 percent). This threshold needs to take into account the characteristics of different businesses (cyclicality) and their ability to service debt. Thresholds for nonallowance of interest expense deductions might be phased in (progressively lowered) over a two- to three-year period.

Allow conversions of debt to equity. This technique, presently prohibited in East Asian countries such as Thailand and Indonesia, should be available outside court-supervised reorganizations or liquidations. To allow it as part of voluntary reorganizations, it would be necessary to amend company laws to permit debt-to-equity conversions for both public limited companies and private limited companies.

Liberalize the legal framework for foreign investment. In Thailand, the Alien Business Law is an impediment to strategic foreign investment that could facilitate the restructuring of Thai corporations. At the same time, investors may circumvent this law by using nominees and shell corporations, which increases costs and risks for investors while failing to protect the interests for which the law was designed. Hence, the government could consider reforms to promote strategic foreign investment, protect Thai national interests, and provide greater transparency in direct foreign investment. Similarly, in Indonesia, a shorter negative list on foreign investment could be put into place.

Liberalize the legal framework for property ownership. Restrictions on foreign ownership of nonagricultural land are an impediment to foreign ownership of commercial real estate, residential property, and manufacturing facilities. At greater cost and risk, foreign investors may circumvent property ownership laws through the use of nominees and heavily taxed long-term leases. Current problems in the financial sector in Thailand largely were caused by an oversupply of residential and commercial property. Additional opportunities for foreign ownership would reduce this oversupply. Hence, some liberalization of property laws could achieve a better balance between supply and demand of developed real estate and promote strategic foreign investment while protecting national Thai interests and providing greater transparency in real estate transactions.

Source: World Bank.

internal control, unreliable financial reporting, lack of adequate disclosures, lax enforcement to ensure compliance, and poor audits. These problems are evidenced by unreported losses and understated liabilities. Regulators responsible for monitoring and overseeing such practices failed to detect weaknesses and take timely corrective action.

Improving the framework for corporate governance and financing takes time and requires considerable behavioral changes. Only countries that have gone through extreme crises have been able to quickly alter governance and distribution of control of the real and financial sector.[8] Although investors have neglected corporate governance, since the onset of the crisis, they have become aware and sensitive to the need of reform. Most East Asian countries have embarked on reforms in their corporate governance (see box 4.6). There are six areas of specific importance to change the corporate governance in East Asia, which are described below.

Enhance enterprise monitoring. The role of commercial banks in enterprise monitoring and corporate governance will have to be enhanced through a comprehensive program of bank restructuring and institutional development. Banks, which in the short-run will dominate East Asian financial sectors, need to become more effective monitors of firms' management in an ex-ante, interim, and ex-post sense. At the same time, banks need to develop an arms-length relationship with corporations. This will require stricter enforcement of limits on lending to connected firms and insiders, the violation of which has contributed to the recent financial crisis and poor intermediation. In those cases where banks and firms are effectively controlled by the same shareholders, increased transparency is required, which could take the form of more disclosure or the requirement of a formal ownership relationship, such as through a holding company. Other financial institutions and agents involved in disciplining firms should be encouraged to enhance their role. For example, bond investors can play an important role in disciplining managers, but this requires some changes in relevant commercial codes.

Improve disclosure and accounting practices. Although disclosure and accounting rules are becoming increasingly consistent with international standards in East Asian countries, the application of these rules appears to be hindered by the limited role of self-regulatory agencies (SROs) in raising standards and practices and imposing sanctions on irregular behavior. A larger role for SROs that is backed by increased legal powers to discipline violators may be needed. In addition, the market structure of the accounting industry, with limited participation by foreigners, may have been a hindrance to upgrading practices. Improved reliability and integrity of financial reporting and disclosures is the prerequisite for reforming the corporate sector and restoring market confidence. Corporate management must exercise its responsibility for preparing financial statements that are transparent and in accordance with standards that are easily understood, not only by the management, but also by regulators and the general public. Introducing truly transparent accounting and auditing systems, consistent with international best practice, would require (a) reducing the role of the government in regulating and overseeing the accounting and overseeing the accounting and auditing practices and profession; (b) establishing an independent and self-regulating national professional body for setting accounting standards; (c) strengthening the financial oversight functions of boards of directors and improving the effectiveness of audit in listed companies by establishing audit committees of boards of directors.

Strengthen the enforcement of corporate governance regulations. The formal corporate governance framework in most East Asian countries is not different from the standards used by developing countries with similar income levels. But the practice and enforcement of corporate governance in East Asian countries are weak. Important changes in the capital markets as well as in the judicial system are needed such that minority shareholders' rights are better protected. The main lead for improvements will have to come from stock market watchdogs. Extra tools to enforce regulations and discipline members may be needed to make these improvements more effective. It may be useful to review the process for appointing commissioners and board members of the stock exchange monitoring bodies.

Improve the corporate governance framework. In the more medium-term, a number of improvements in the corporate governance framework are desirable. For example, the 1997 proposal by the Stock Exchange of Thailand for self-regulation on corporate governance of listed firms could be made mandatory (the proposal was to adopt standards regarding the roles, duties, and responsibilities of the directors of listed companies). Generally, countries could benefit from a broad public discussion on the topic of corporate governance, similar to what happened in the United Kingdom and other developed countries in recent years.[9] In the end, the issue of corporate governance concerns the distribution of control in the economy over the real sector. A discussion of the preferred evolution of the real or industrial sector should form the basis of the desired evolution of the corporate governance framework. The process of consultations used for the 1998 Organization for Economic Co-Operation and Development (OECD) report on corporate governance provides a good starting point on how this discussion might be conducted.

Facilitate equity institutions. As external financing needs are high, particularly for new equity, attracting

World Bank support of changing corporate governance

Along with its loans, the World Bank provides technical advice on policy, often expressed as agreed conditions in loans. Here's a sampling:

In Thailand, a US$350 million Finance Companies Restructuring Loan in December 1997 helped conduct in-depth assessments of the financial condition of the nonsuspended finance companies and helped rehabilitate these institutions. The loan also helped strengthen prudential regulation and the supervisory regime. The Economic and Financial Adjustment Loan helps with financial sector reform and corporate recovery.

In Indonesia, the Bank provided a US$1 billion Policy Reform Support Loan in July 1998 to help clarify governance and supervision of banks and to set up an Indonesian Bank Restructuring Agency.

The March 1998 US$2 billion Structural Adjustment Loan to Korea, following up on actions promised under the US$3 billion Economic Reconstruction Loan of December 1997, supported improvements in the governance structure of banks. The loan was also designed to help the government improve the reliability of key financial information provided by banks and corporations to regulators, shareholders, and the general public; promote effective monitoring of corporate performance by boards of directors and shareholders; and facilitate efficient liquidation of insolvent corporations.

In Malaysia, the June 1998 US$300 million Economic Recovery and Social Sector Loan had the objective of improving accounting standards of the International Accounting Standards Committee and suspending of stockbroker companies for failure to comply with capital adequacy standards.

Source: World Bank loan documents.

---

new investors will be important. To facilitate the process of new equity infusions, it will be necessary to provide new investors with a more direct role in monitoring and disciplining managers. This will require a representation of minority shareholders on the board of directors, which in turn may ensure broader application of the one-share, one-vote principle and use of cumulative voting for the appointment of directors. It also may be useful to introduce supermajority voting rules for fundamental corporate decisions, such as acquisitions and major investments. Some market participants and analysts even have suggested that new equity infusions may require a more-than proportional representation on the board of directors by new equity owners, at least until other investor protection mechanisms are strengthened. Improving corporate regulations will require enhancing the role of institutional investors in monitoring firms, which will have to begin with improving the regulations of the investors.

Strengthen institutions. In terms of institutional development, it is clear that data availability and analysis of corporate financing and governance represent major weaknesses in East Asian countries. Not only are the data on corporations, especially on small and medium enterprises (SMEs), incomplete and of poor quality, there are also institutional gaps as the responsibility for monitoring firm performance and behavior is scattered. Follow-up work should aim at systematizing data collection on firms and performing more and regular surveys, which should be a joint effort of private, semi-public, and public organizations.

Notes

1. All computations in this section, unless otherwise stated, represent the median firm based on the Financial Times Extel and Worldscope databases, which include all firms listed on the stock exchange in their respective countries.

2. Based on simple Pearson correlation tests.

3. One particular aspect of concern are disclosure rules. Regarding what to disclose in developing countries, whereas most developed markets rely on market practice and due diligence obligations to ensure disclosure of all material information, it is prudent for the authorities to be more proactive. In several East Asian countries, however, markets still were struggling to define precisely what this meant in practical terms. Disclosure systems were also weak in how information was disseminated through public repositories and mandated requirements for publicly held firms. This weakened market incentives, particularly for financial intermediaries and for firms issuing short-term paper.

4. Morck, Shleifer, and Vishny, 1988.

5. The impact of two financial shocks faced by corporations in each of the five crisis countries is considered: (a) the increase in financial obligations implicit in the devaluation of the exchange rate, and (b) the increase in financial outlays caused by rising borrowing rates. For each country, the first shock was approximated as the increase in the value of foreign debt in domestic currency determined by the devaluation recorded on average over the first two weeks of September of

1998 with respect to the precrisis (March 1997) value of the U.S. dollar (in revising ROA, only one-third of this shock was deducted from actual net income). The second shock was approximated by the increase in lending rates in the first few months of 1998 with respect to their level observed in the equivalent period of 1997. These shocks were then applied to the balance sheets of each firm as of end 1996. Thus, a "1996 with shock" is obtained and compared with the "1996 actual." Because many of the *ceteris paribus* assumptions made for simplicity have changed, the "1996 with shock" is obviously just a reasonable first-cut approximation to what may be happening. The results are provided for the median firm based on a sample of nonfinancial firms listed on the stock exchange in the respective countries. For more detail see Claessens, Djankov, and Ferri (1998).

6. Dress, B. and C. Pazarbasioglu, *The Nordic Banking Crisis, Pitfalls in Financial Liberalization?* IMF Occasional Paper 161, April 1998.

7. The scheme is modeled on the Fideicomiso para la Cobertura de Riesgos Camiarios (FICORCA) plan to facilitate restructuring of Mexican private corporate external debt in the aftermath of the 1982 debt crisis.

8. Chile is an example of a country that achieved significant ownership and control transformation of its economy following its financial crisis of the early 1980s. The transformation involved a reduced role for conglomerates, the privatization of state enterprises, a fully funded pension system, and various other tools. Many transition economies also have been able to achieve a rapid transformation.

9. For example, the Cadbury (1992) and Hamel (U.K., 1998) report, the Toronto Stock Exchange (Canada, 1994) report, the Peters report (Netherlands, 1997), the Corporate Governance Forum (Japan, 1997), the Statement on Corporate Governance (U.S., 1997), and similar efforts in a number of other countries.

chapter five

From Economic Crisis to Social Crisis

*It was the rich who benefited from the boom...but we, the poor, pay the price of the crisis. Even our limited access to schools and health is now beginning to disappear. We fear for our children's future" said Khun Bunjan, a community leader from the slums of Khon Kaen, northeast Thailand and her husband, Khun Wichai. Khun Wichai recently lost his job in the local factory and his wife is selling less at the local market. As a result they took both their son and daughter out of school and put them to work. "What is the justice in having to send our children to the garbage site every day to support the family?" questions Khun Bunjan. But Khun Wichai thinks he is lucky. His neighbors are sending their children to beg and some girls became prostitutes. Among the older male youths, drug dealing has become an increasingly attractive source of income. Increased competition for survival, frustration and psychological stress are all leading to heightened household and community tension. This tension has led to increased domestic violence and with fewer jobs, neighbors who once cooperated are now competing. Stealing, crime, and violence are on the rise. People are feeling unsafe and insecure. "This breakdown of our community's networks will affect stability," added Khun Bunjan. —*World Bank staff interviews.

East Asia's economic and social structures are under strain, and decades of unparalleled social progress risk being undone. In country after country growth is declining—from historical average per capita increases of over 5 percent per annum to negative levels over the next year—with especially sharp contractions expected in Indonesia, Thailand, and the Republic of Korea. The poor are being hit hardest during the crisis as demand for their labor falls, prices for essential commodities rise, social services are cut, and crop failures occur in countries experiencing drought. The combined macroeconomic and agricultural shock undercuts fragile coping mechanisms, especially in Indonesia, and could be life-threatening if consumption levels drop sharply. Widespread economic hardships are tearing at the fabric of society: food riots and ethnic tensions in Indonesia, farmers are protesting in Thailand, and workers are voicing discontent in Korea. These signs of stress on social systems are politically worrisome. Moreover, the increasing tensions at both the household and community level are equally damaging, especially for women and children. Children are being pulled out of school and put to work; food is being rationed within the household, and women and girls are frequently the first to sacrifice their portions; and violence, street children, and prostitution are all on the increase. This is the human crisis.

A rapid return to macroeconomic stability and growth through distributionally favorable adjustment policies is the only way to begin to put a floor under the falling incomes of the poor. In the meantime, government action can curb the welfare losses of the poor in the short-term, and help protect their human resource investments. It is essential to ensure that food markets work, to augment the purchasing power of vulnerable households, to cushion the impact of price increases, and to preserve the poor's access to health and education. Strengthening public and private institutions responsible for service delivery is also crucial in both the short- and longer-term.

East Asian countries encountered important social challenges before the crisis. In most of East Asia, social policies developed against the backdrop of political stability, full employment, high household savings, and relatively strong community ties giving governments little reason to plan for downside risks. But even during times of economic growth, three issues challenged social policies in East Asia: persistent pockets of poverty and rising inequality, outmoded labor market policies and industrial relations, and rising needs for formal mechanisms to support household security. Growth masked those problems but when the crisis stripped this mask away, the region's persistent social vulnerabilities were sharply revealed.

Progress and vulnerabilities

East Asian countries have achieved spectacular welfare gains in the last two decades.[1] Consistently high growth rates have been translated into quantifiable welfare improvements primarily because growth has largely been inclusive. Public provisioning of social services has been widespread and the productivity of the poor and their employment opportunities have increased enormously. The number of poor has fallen and the severity of poverty has declined. Life expectancy at birth, infant mortality, and literacy have all improved. These achievements are even more impressive when compared with social developments in other regions or other developed countries during their decades of industrialization.

Impressive measures of social development

Between 1975 and 1995, poverty in East Asia dropped by two-thirds according to the region's head-count index using the constant US$1-a-day poverty line (in 1985 purchasing power parity [PPP] terms). This pace of poverty reduction was faster than in any other developing region. In 1975, six out of ten East Asians lived in absolute poverty according to this standard; by 1995, the ratio had dropped to two out of ten (see table 5.1). This means that the number of poor people in the region was more than halved, from 720 million to 345 million. Further, the rate of decline accelerated after 1985. The number of people in poverty fell by 27 percent between 1975 and 1985; the decline was 34 percent between 1985 and 1995.

Changes in poverty levels and rates varied across the region. In 1975, 92 percent of the region's poor lived in China and Indonesia, primarily because they were the two most populous countries. Since then, however, both countries have seen substantial declines in poverty: by 82 percent in Indonesia and 63 percent in China. In absolute terms, the number of poor was more than halved in China, and fell by almost three-fourths in Indonesia. As a result, by 1995, the two countries' share of the region's poor had dropped to 84 percent. Although Indonesia's record is remarkable—the head-count declined from 64 percent in 1975 to 11 percent in 1995—Thailand had the largest proportional reduction

TABLE 5.1
Poverty in East Asia, summary statistics: 1975–95

Economy	Number of people in poverty (million)			Head-count index (percent)			Poverty gap (percent)		
	1975	1985	1995	1975	1985	1995	1975	1985	1995
East Asia[a]	**716.8**	**524.2**	**345.7**	**57.6**	**37.3**	**21.2**	**n.a.**	**10.9**	**6.4**
East Asia (exc. China)	147.9	125.9	76.4	51.4	35.6	18.2	n.a.	11.1	4.6
Malaysia	2.1	1.7	0.9	17.4	10.8	4.3	5.4	2.5	< 1.0
Thailand	3.4	5.1	< 0.5	8.1	10.0	< 1.0	1.2	1.5	< 1.0
Indonesia	87.2	52.8	21.9	64.3	32.2	11.4	23.7	8.5	1.7
China	568.9[b]	398.3	269.3	59.5[b]	37.9	22.2	n.a.	10.9	7.0
Philippines	15.4	17.7	17.6	35.7	32.4	25.5	10.6	9.2	6.5
Papua New Guinea	n.a.	0.5	1.0[c]	n.a.	15.7	21.7c	n.a.	3.7	5.6[c]
Lao PDR[d]	n.a.	2.2	2.0	n.a.	61.1	41.4	n.a.	18.0	9.5
Vietnam	n.a.	44.3[e]	31.3	n.a.	74.0[e]	42.2	n.a.	28.0[e]	11.9
Mongolia	n.a.	1.6	1.9	n.a.	85.0	81.4	n.a.	42.5	38.6

na: not available.
Notes: All numbers in this table (except for Lao People's Democratic Republic) are based on the international poverty line of US$1-a-day per person at 1985 prices.
a. Includes only those countries presented in the table.
b. Data relates to 1978 and applies to rural China only (World Bank 1996d).
c. Data relates to 1996.
d. Available data on PPP exchange rates and various price deflators for Lao People's Democratic Republic (LAO PDR) are not very reliable and lead to anomalous results. The poverty numbers for Lao PDR in this table are based on a national poverty line which is based on the level of food consumption that yields an energy level of 2,100 calories per person per day and a non-food component equivalent to the value of non-food spending by households which are just capable of meeting their food requirements. The US$1-a-day poverty line is based on characteristic poverty lines in low income countries that have comparable basis in food and non-food consumption needs; the poverty numbers for Lao PDR are, therefore not strictly comparable to those for other countries.
e. The figures refer to 1984. "Household Welfare in Vietnam's Transition" in Macroeconomic Reform and Poverty Reduction, edited by D. Dollar, J. Litvack, and P. Glewwe. World Bank Regional and Sectoral Study, 1998.
Source: *Everyone's Miracle?*, World Bank, 1997.

TABLE 5.2
Social indicators in East Asia: 1985–95

Country	Life expectancy at birth		Infant mortality rate (per 1,000 live births)		Primary net enrollment (percent)		Secondary net enrollment (percent)	
	1970	1995	1970	1995	1970	1995	1970	1995
East Asia	**59.4**	**68.8**	**76**	**34**	**na**	**na**	**na**	**na**
Taiwan (China)	69.0	74.8	69	6	na	> 99	75.0	87.4
Korea, Rep. of	60.6	72.0	46	10	> 99	> 99	45.4	93.4
Malaysia	61.6	71.8	45	12	84.1	88.7	25.5	55.9
Thailand	58.4	69.0	73	35	78.6	88.2	18.2	34.9
Indonesia	47.9	63.7	118	51	75.6	> 99	13.0	55.0
China	61.7	69.4	69	34	75.9	> 99	34.7	50.7
Philippines	57.2	66.5	71	39	> 99	> 99	40.4	75.5
Papua New Guinea	46.7	58.5	112	68	30.8	70.0	3.7	13.3
Lao PDR	40.4	52.8	146	104	na	60.0[b]	na	15.0[b, c]
Vietnam	49.36	67.5	111	42	na	91.0[d]	na	45.0[c, d]
Mongolia	52.70	66.4	102	55	na	na	na	na

na: Not available.
Notes: a: Source: ROC, various years.
b. 1993, Source: World Bank, 1995a.
c. Lower secondary.
d. Source: World Bank, 1996.
Source: Ahuja and Filmer (1996), for net enrollment rates; World Bank data for life expectancy and infant mortality.

between 1975 and 1995, from 8 percent to less than 1 percent.

Between 1973 and 1990, the region saw substantial gains in life expectancy and declines in infant mortality (see table 5.2). Similarly, access to education expanded—China and Indonesia reportedly joined Korea and the Philippines in achieving universal primary net enrollments. In those four countries, plus Malaysia, secondary net enrollment also expanded beyond 50 percent of children in the eligible age group.

Five major factors contributed to the region's social progress and inclusive development.[2] Some could be jeopardized by the crisis.

- *Small holder-based rural development.* In most East Asian countries, small-scale family farming has dominated agricultural production, and government policies have contributed to an equitable development path by supporting productivity growth on family farms though infrastructure (notably in irrigation and roads), appropriate pricing and other market policies, and new technologies.

- *Rapid growth in demand for non-agricultural labor.* In all of the region's major economies, except the Philippines, the past few decades have witnessed a massive labor shift—out of agriculture into more productive work such as rural non-farm activities, urban industry, and services. This shift was driven by many factors: agricultural incomes rose, spurring demand for rural non-farm employment, internal economic integration increased, overall capital deepened, and industrial employment grew rapidly. In Malaysia, the share of wage workers in industry and services rose from 30 percent in 1960 to more than 60 percent in the 1990s.[3] Since the early 1980s, rural, and then urban industry and service employment, has expanded enormously in China. Such changes have been accompanied by large increases in real wages—and by large increases in the incomes of self-employed workers.

- *Widespread public provision of basic education and health services.* Rapid growth in the public provision of schooling was a major element of the region's human resource strategy. Primary schooling expanded rapidly, followed by growth in secondary and then tertiary education. These efforts contributed to the early achievement of almost universal primary education, complemented by a significant expansion of basic health services, including key preventive services such as immunizations and basic curative care.

- *"Flexible" labor markets and low labor market dualism.* East Asian labor markets are fairly flexible, with fewer institutional or policy-driven rigidities than European or Latin American markets—minimum wage policies are limited, wage-setting practices are flexible, and wages and productivity growth are closely linked. As a result, fewer sharp contrasts existed between formal, privileged workers and rural, informal workers.

- *Upgrading work force skills and investing in education ahead of demand.* In Korea, education expansion effectively anticipated the changing demands of modern industries and services. Young people with a primary school education were the core of the work force for the early phases of labor-intensive industrialization. As productivity and wages rose—driven by high levels of capital investment and technological advances—demands shifted first to secondary and increasingly to tertiary graduates. Meanwhile, schooling expanded rapidly enough that skills supplies actually surpassed demands. Between the early 1970s and the late 1980s, this prompted a decline in wage inequality among workers with college, secondary, and primary schooling.[4] While young school graduates tended to enter and remain in industry, older workers were moving out of agriculture directly into services. Rural productivity rose rapidly and the unskilled labor market tightened, prompting a convergence of rural and urban incomes.[5]

It would be easy to overstate the factors which supported social and economic growth. For example, Thailand's sharp rise in inequality appears to be partly related to slow growth in secondary education; China's unusually small tertiary education sector could cause future problems. In Indonesia and the Philippines, there are concerns about education quality. Some East Asian countries have done very well in raising people's overall level of health, but, by some measures, unusually high levels of child mortality still exist in Indonesia, Korea, and the Philippines.[6] In some cases, labor market policies have been overly restrictive, especially in Korea and China, where insiders are highly protected and state intervention in employment is pervasive.

Some countries strive to avoid insecurity—Korea is following Japan toward a European style, pay-as-you-go, welfare state; China is providing cradle-to-grave protection for state sector employees and their families; Malaysia and Singapore are running large-scale, state-managed provident funds.

Pre-crisis challenges and emerging vulnerabilities

Even before the onset of the crisis, three issues emerged to challenge social policies: protracted poverty and rising inequality; concerns about labor rights; and rising demands for formal mechanisms to offset household insecurity. The crisis has aggravated conditions underlying each of these issues.

Persistent poverty and vulnerability

Despite tremendous gains, poverty continues for many in East Asia. Poverty is still high in Indochina and Mongolia, reflecting slower growth and recent systemic transition. In high growth countries, vulnerabilities also remain considerable given the large numbers of households just above the poverty line. In Indonesia, a 25 percent increase in the poverty line results in more than doubling the head count index, from 11 to 25 percent in 1996. Moreover, absolute poverty persists in certain areas or among certain groups. The poor tend to live in rural areas, have less education and live in households headed by farmers. In addition, some ethnic minorities are disproportionately poor and girls appear to get shortchanged in household resource allocation—particularly in poor households.[7]

Reductions were not uniform across these economies, and poverty remains acute in some regions. For example, in 1990, poverty incidence in Indonesia ranged from 1.3 percent in Jakarta to 46 percent in West Nusa Tenggara. In the inland province of Guizhou in China, the incidence of poverty in 1992 was 20 times that in the booming coastal province of Guangdong. In Thailand, the northeast has the highest incidence of poverty and the highest concentration of poor people. Even in Vietnam and Lao PDR, where poverty is much more widespread, regional variation is substantial. In Vietnam the incidence of poverty ranges from 34 per-

cent in the southeast to 77 percent in the north-central region.

Are some regions poor because they contain a high concentration of households with characteristics that are strong indicators of poverty? Or are there purely geographic effects? Jalan and Ravallion (1997) suggest that in China geographic effects remain strong even after controlling for household characteristics: households with identical characteristics experience different or even diverging consumption growth depending on location.[8] This finding suggests that policies to augment geographic and community capital are essential to help alleviate poverty.

Rising inequality

East Asia has been credited with achieving "growth with equity" but the facts are more varied. On average, East Asia's income distribution has remained largely unchanged in the last 15 years, either in absolute terms or relative to other regions. An aggregate index of inequality finds that East Asia is more egalitarian than Latin America or Sub-Saharan Africa but less egalitarian than former socialist countries in Eastern Europe, high-income countries, and South Asia (see table 5.3).

TABLE 5.3
An international comparison of inequality
(averaged Gini coefficients)

Region	1980s	1990s
Eastern Europe	25.0	28.9
High-income countries	33.2	33.8
South Asia	35.0	31.9
East Asia and the Pacific	38.7	38.1
Middle East and North Africa	40.5	38.0
Sub-Saharan Africa	43.7	47.0
Latin America and the Caribbean	49.8	49.3

Note: The total sample includes 108 economies. Although Gini coefficients come from household surveys that satisfy comprehensiveness criteria in terms of both geographical coverage and income sources, they nevertheless include unadjusted data from both expenditure and income distributions. The proportion of income Gini coefficients vary across regions, hampering comparability. Regional averages are unweighted, and changes across the two decades may be due to changes in the composition of the sample. The numbers merely suggest broad orders of magnitude.
Source: Deininger and Squire (1996).

Within East Asia there was significant variation among countries (see table 5.4).[9] Inequality has clearly risen in China, Hong Kong (China), and Thailand. It also appears to have inched up in the Philippines between 1985 and 1994, and recent data for 1997 suggest a sharper increase.[10] Only Malaysia shows a slight decline in inequality, although this reflects earlier gains which were partially reversed by a significant rise in inequality in the 1990s. With the exception of China, all four countries now have inequality rates well above the regional average.

TABLE 5.4
Inequality in East Asia

Economy	Period	Measured variable	Gini coefficient (percentage points)	
			First Year	Last Year
Hong Kong (China)	1971–91	I/H	40.9	45.0
Singapore	1973–89	I/H	41.0	39.0
Taiwan (China)	1985–95	I/P	29.0	31.7
Korea, Rep. of	1970–88	I/H	33.3	33.6
Malaysia	1973–95	I/P	50.1	48.5
Thailand[a]	1975–92	E/P	36.4	46.2
Indonesia	1970–95	E/P	34.9	34.2
China[b]	1985–95	I/P	29.9	38.8
Philippines	1985–94	E/P	41.0	42.9
Papua New Guinea	1996	E/P		50.9
Lao PDR	1993	E/P		30.4
Vietnam	1993	E/P		35.4
Mongolia	1995	E/P		33.2

Note: I/P is per capita income, E/P is per capita expenditure, and I/H is income per household. The numbers in this table may be marginally different than those reported in other World Bank reports based on unit record data. For the sake of consistency across countries, we only report Ginis based on grouped data, except for Korea, Singapore, and Hong Kong, China, which are from Deininger and Squire 1996.
a. Thailand is the only country for which we can present Ginis based on both expenditure and income distribution. The per capita income-based Gini (I/P) was 42.6 percent in 1975 and 54.6 percent in 1992.
b. Because of China's size, as well as with valuing home production of grain for own consumption, controlling for spatial price variations, and valuing in-kind transfers, the uncertainty associated with Chinese Ginis may be even greater than that for other economies (see World Bank 1997c for a detailed discussion).
Source: Deininger and Squire (1996), and World Bank staff calculations.

Research on China and Thailand suggests two explanations for rising inequality. First, the returns to higher levels of education have increased, which is driving a wedge between highly skilled workers and those with primary or lower-secondary education. Second, spatial disparity in economic prosperity is growing because activity is concentrated in certain areas. The current crisis may increase inequality in access to education and also affect areas within the region differently, further aggravating skills- and geography-based differentials.

High inequality negatively effects society in three dimensions: undermining poverty alleviation, impeding growth, and contributing to social tension. For a given growth rate, an increase in inequality tends to lead to slower poverty reduction.[11] Greater initial inequality can reduce economic growth because of imperfect

credit markets or political economy channels. Social tension can result when the benefits of growth accrue unequally to easily identifiable groups—for example, certain regions, ethnic groups, or men and boys—even if these are not major factors in overall inequality. Japan, Korea, and China are notably homogenous societies, but Indonesia and Malaysia have major divides across ethnic lines that have spilled over into significant conflicts in the past—into the bloodbath of 1965 in Indonesia and periodic violence against Chinese-Indonesians since. Yet, both countries achieved sufficient social understanding and stability to foster high levels of capital investment, notably by ethnic Chinese groups. In Malaysia's case, there was a highly managed process of affirmative action for the ethnic Malay population, or Bumiputra (see box 5.1). Managing inequality is a challenge both for rich and poor countries, as they attempt to balance incentives for superior individual performance with acceptable levels of inequality which foster poverty reduction and growth.

BOX 5.1
Malaysia's NEP and social inequality

Ethnic rioting came to a head in Kuala Lumpur in May 1969 after the Chinese-dominated opposition party won many parliamentary seats from the Malay-dominated ruling coalition. Malays, most of whom were impoverished, reacted violently in fear of losing their political influence. To confront and combat the causes for this uprising, the government launched the New Economic Policy (NEP), an affirmative action plan designed to pull poor Malays into the mainstream of the country's economic system.

The NEP introduced a series of government regulations, quotas, scholarships, and other privileges designed to help Malays. The results have been impressive: Malay's share of national wealth jumped to 20.6 percent in 1995 from 2.3 percent in 1970. Much of the NEP's success is attributed to education as the number of Malay doctors, lawyers, and engineers increased dramatically. Malaysia's affirmative action policies have not been without controversy. Some have observed that the policies helped foster favoritism and inefficiency. Even so, the overall approach clearly succeeded in supporting high levels of investment, low levels of internal conflict, and rapid advancement of all Malaysians, especially the Bumiputra, for more than two and a half decades.

Source: Murray Heibert, "Lessons From Malaysia," *Far Eastern Economic Review,* May 28, 1998.

Labor relations

Most countries in the region sought to maintain relatively unfettered labor markets during the early stages of economic development at the expense of granting workers the right to bargain collectively. Until the late 1980s, labor conditions in Korea, Malaysia, the Philippines, Taiwan, China, and Thailand were determined unilaterally by employers with or without the assistance of government.[12] However, in the late 1980s changes began to affect unions with shifts toward more democratic governments, pressures for modern industrial relations brought about by tighter labor markets, and more sophisticated production processes.

Failure to modernize worker-management relations in countries with sophisticated economic and political structures can become costly, as Korea's experience since the late 1980s has shown. If there is no industrial relations system to allow workers to air grievances and resolve disputes, strikes and other forms of job actions, sometimes violent, can become common. One of the challenges facing maturing economies in East Asia is to manage industrial relations to protect workers' legitimate rights, while avoiding granting entitlements that result in resource inefficiencies.

Household risks

Most East Asian households have few formal mechanisms to protect them from risks associated with job losses, disabilities, and aging. Instead, most rely primarily on personal savings and informal family and community links. A few countries in the region have set up formal schemes to address household insecurity, but these cover small portions of the population (state workers in China, large enterprises in Korea, participants in the state provident funds in Malaysia and Singapore). The current financial crisis makes glaringly evident the absence of formal provisioning for household security.

In the face of the crisis, the demands for safety nets are urgent, but they will be far more dramatic 25 years from now because East Asian societies are undergoing a rapid demographic shift. East Asian populations are aging, moving into cities, and increasingly working in

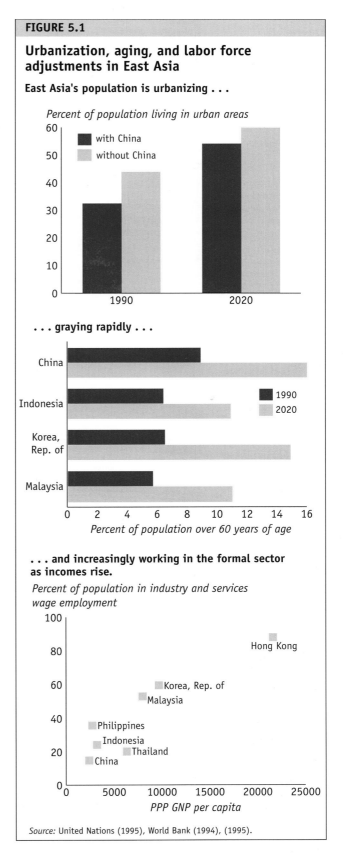

FIGURE 5.1

Urbanization, aging, and labor force adjustments in East Asia

East Asia's population is urbanizing . . .

Percent of population living in urban areas

(with China / without China; 1990, 2020)

. . . graying rapidly . . .

(China, Indonesia, Korea, Rep. of, Malaysia; 1990, 2020)

Percent of population over 60 years of age

. . . and increasingly working in the formal sector as incomes rise.

Percent of population in industry and services wage employment

(Hong Kong, Korea, Rep. of, Malaysia, Philippines, Indonesia, Thailand, China)

PPP GNP per capita

Source: United Nations (1995), World Bank (1994), (1995).

the formal sector (see figure 5.1). In France it took 140 years for the proportion of the population over 60 to double—from 9 to 18 percent. By contrast, in Korea, people over 60 will double their share in the population in only 30 years—between 1990 and 2020—and in China the portion of people over 60 will rise from 9 percent to 16 percent of the population during the same period.[13] These changes reflect a swifter demographic transition, and in China, they demonstrate the combined effects of early gains in rural health status and an activist population policy. All three trends will strain informal family-based mechanisms of household protection, and will increase demands for formal, government-mandated schemes.

The social impact of the crisis

The economic contraction is affecting the lives of millions, and aggravating social vulnerabilities. It is likely to have many dimensions—falling incomes, rising absolute poverty and malnutrition, declining public services, threats to educational and health status, increased pressure on women, and increased crime and violence. In Indonesia there is also a radical breakdown in social order as an increasingly fragile social equilibrium was brought under intolerable stress by the collapse in economic confidence and fall in incomes.

The effects of the crisis are acute in Indonesia, and severe in Thailand, Korea, and Malaysia. The Philippines has been less affected, but also shows signs of worsening social conditions. After declining steadily for five years, in September 1997 there was a *rise* in self-reported poverty.[14] Trade, capital flow, and migration linkages among countries are hastening the transmission of economic and social effects across the region. While China remains largely insulated, falling regional demand and slowing intra-regional foreign investment are aggravating domestic difficulties. Countries in Indochina are experiencing growth downturns and financial difficulties as the impact of the regional crisis unfolds with a depth and intensity far exceeding previous expectations. The Pacific Islands have also been hit; by far the worst affected is the Solomon Islands where gross domestic product (GDP) is expected to shrink by 10–12 percent, driven by a collapsing log export market in which export prices have halved.

Effect on households

The economic crisis is having four severe effects on households: falling labor demand, sharp price shifts, a public spending squeeze, and erosion of the social fabric. In addition, some countries have been simultaneously hit by drought.

Falling labor demand. Economic decline, the corporate crisis, and a credit squeeze are causing lay-offs, real wage declines, weak demand for new labor market entrants, and falling margins in the informal sector. Whether the impact in a given country is primarily through higher unemployment or lower wages depends on societal and economic structures. In Thailand, unemployment has increased by 50 percent since the start of the crisis to 1.5 million in February 1998, and is expected to exceed 6.0 percent by year-end. In Korea, unemployment reached 7.0 percent in June 1998 and may affect as many as 2 million people during 1998, up from 0.5 million in 1997 (see figure 5.2). In the Philippines, 1 million additional people joined the ranks of the jobless between April 1997 and April 1998, raising the unemployment rate to 13.3 percent. In Indonesia, where some 4.5 million people (4.9 percent) were already unemployed in 1996, official estimates suggest an additional 10 million may lose their jobs by early 1999, although it is likely that many of them will move into low-paying urban and rural informal sector work.

This is *not* essentially an urban shock, despite the high profile of urban unemployment figures. Rural areas will also be seriously affected by labor movements, production linkages, and intra-household relationships because of the highly integrated nature of the urban and rural economies and the declining demand in urban areas. Increased *under*-employment and falling wages may be more widespread and valid indicators of a decline in well-being than unemployment statistics. In Korea, unemployment has increased sharply; however, there has also been an increase in the number of family and agricultural workers and in the number of labor force drop outs, suggesting substantial increases in underemployment. Data from western Java, in Indonesia, point to a decline in rural real wages by 10 percent between August and December 1997. Available information in both Indonesia and Thailand suggests that workers are returning to their villages from the

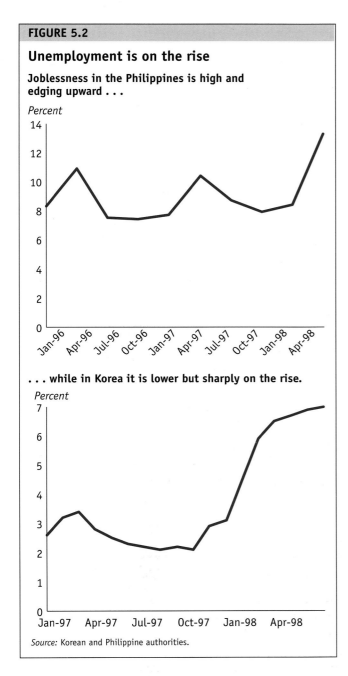

FIGURE 5.2

Unemployment is on the rise

Joblessness in the Philippines is high and edging upward . . .

Percent

. . . while in Korea it is lower but sharply on the rise.

Percent

Source: Korean and Philippine authorities.

urban centers of Jakarta and Bangkok. There is an international dimension as well: migrant workers in Malaysia, notably those from Indonesia, are also losing their jobs or are suffering real wage declines. The Philippines is also vulnerable because of large numbers of overseas workers. Finally, there is evidence that women are disproportionately targeted for lay-offs (see box 5.2).

Sharp price shifts. Prices, particularly those of basic necessities such as food and medicine, have risen dramatically because of exchange rate devaluations. In Indonesia, where the exchange rate has depreciated by 80 percent since July 1997, the prices of antibiotics doubled between October 1997 and March 1998 and the consumer price index (CPI) for food increased by more than 50 percent between June 1997 and March 1998, compared with a 38 percent rise for the general CPI. Throughout the region, prices of drugs have risen sharply—most dramatically in Indonesia, where there are already reports of households postponing vaccinations or use of other drugs. The high cost of medication also makes HIV/AIDS patients in the region particularly vulnerable. The longer-term effect of these price changes is difficult to predict because in most countries the exchange rate appears to have overshot an equilibrium position, and it is hard to know how much will be inflated away or reversed by increased demand for local assets once confidence recovers. But there will clearly be short-run price effects. These will squeeze profits in much of the informal sector, which is, in general, already facing weak demand in the domestic market.

Public spending squeeze. Public spending is being constrained by revenue falls, the effects of exchange rate changes on interest bills and the need to finance the increasing liabilities of the government in financial sector and corporate restructuring. Budgets were cut initially in all of the affected countries as part of the macroeconomic adjustment process. While fiscal targets have since been adjusted in some countries, public services may still suffer cutbacks, creating both short- and long-term impacts on households. Of particular concern are potentially irreversible effects of cutbacks on investment in human resources. Budget constraints may also mean funding is cut in specific areas which benefit relatively poor households (such as small holder treecrop investment).

Erosion of the social fabric. Economic stresses are leading to social and political problems. The rapid development which brought rising incomes to most East Asian households in the past decades, has also led to rapid social change, urbanization, migration, and expansion of education. The sudden stop to this rapid growth is expected to disturb the social equilibrium. Social unrest in Indonesia vividly showed how fragile this equilibrium was, and how quickly socially repressed societies can be thrown into upheaval once civil society begins to question the central source of political legitimacy, leadership's ability to deliver continued improvements in economic well-being. On a less dramatic but nevertheless significant scale, social stresses are mounting at both the household and community levels in all crisis countries. Focus group discussions suggest that households with falling incomes are coping by increasing the workload of mothers and by taking children out of school and putting them to work.[15] Economic stress brought on by the crisis may also be leading to increased domestic and community violence and illegal activities such as prostitution and drug trade.

Drought. Parts of Indonesia, the Philippines, and Thailand have been hit by an El Niño-induced drought.

While the effects appear to be less serious than originally feared, in some areas the combined macroeconomic and agricultural shock makes coping with the drought particularly difficult. Consumption drops can be life-threatening, especially in Indonesia where distribution channels and markets have been damaged and foods stocks have dwindled. In the Philippines, the large increase in unemployment between April 1997 and April 1998 was entirely due to the sharp drop in agricultural employment as a result of El Niño, partic-

ularly among unpaid family workers and independent farmers. Households that have suffered from repeated droughts for the past several years—for example, in Indonesia's Eastern Islands—are particularly vulnerable as they are likely to have used up stores of assets and stocks to keep consumption rates constant. The extent of crop failure varies widely and food price rises are magnifying these distributional effects. While higher prices help farm households as they are able to raise surplus food to sell, they also hurt long-term food deficit households and other farm households that are temporarily without food because of the drought.

These shocks will affect incomes, well-being, and access to services, and will interface with the coping strategies that households adopt to protect their consumption levels (see box 5.3). The effects on the poor will depend on the depth and duration of the macroeconomic recession and on whether distribution worsens or improves during the crisis. Both dimensions are uncertain. Other episodes of economic contraction provide limited insights into expected trends in distribution. In many Latin American countries, distribution worsened during the economic difficulties of the 1980s, but in Malaysia's most recent contraction in 1984–87, distribution improved enough to prevent a rise in poverty levels.[16]

Future impacts of the crisis on poverty

Forecasting poverty is hazardous enough in conditions of stability; therefore the following discussion only seeks to illustrate potential outcomes for the poor under alternative assumptions on growth and distribution for four affected Southeast Asian economies.[17] Results are shown based on two commonly used poverty lines: US$1 a day per person (in 1985 purchasing power prices), which is close to poverty lines used in poorer countries (and close to the Indonesia poverty line) and US$2 a day per person, which is closer to national lines used in middle-income countries. In Malaysia and Thailand there are few people living on less than US$1 a day, but as many as 15–20 percent survive on less than US$2 a day. In contrast, half of the populations of Indonesia and the Philippines still live on less than US$2 a day.

The first scenario simulates the effect on poverty of a 10 percent decrease in aggregate consumption or

People are struggling to get by

Entering the informal labor market. Falling household incomes in all countries have already forced many families to send more women, children, and elderly into the labor force. In Thailand, NGOs report an increase in child labor, child prostitution, and child beggars. In the slum settlement of Teparak, Khon Kaen, in northeast Thailand, women were justifiably angry because they had to send their children to the garbage site every day to support the family. In Indonesia, there are reports of children leaving school to join padat karya programs (labor-intensive projects).

Migrating. In Thailand, rising urban unemployment is likely to force industrial workers back to their villages. A Tambon representative, in the rural village of Sap Poo Pan, in northeast Thailand, estimated that out of the village population of 260, 40 people had already returned because of the crisis and 70 were still working outside the village, mainly in Bangkok. Returning laborers will increase the competition for jobs, further marginalizing farm laborers. The impact of migration from neighboring countries is still difficult to define. In Cambodia's northwest province, whole communities relied on jobs in Thailand. Focus groups reported that within the last three months, most of these workers are returning because they have lost their jobs while many remain in debt.

Cutting down on household expenditure. In slum areas, people reported cutting down from three meals per day to two, or even one. In Maluku and South Sulawesi in Indonesia, school principals complained that parents were having difficulty paying parent association fees on time, if at all. In both Indonesia and the Philippines, teachers reported that children were eating less before coming to school in the morning and buying less from vendors and that this was affecting students' ability to concentrate.

Illegal activities. In some slum areas in the Philippines and Thailand, people report an increase in crime. The slum dwellers in Bangkok told focus groups that unemployed youth were already turning to selling drugs to support their families. And, in Cambodia, there were reports of increases in the trafficking of women and children.

Source: Robb, Caroline (1988). "Social Impacts of the East Asia Crisis: Perceptions from Poor Communities." Prepared for East Asian Crisis Workshop, July 1998, IDS, University of Sussex, UK.

income between 1997 and 2000 using a poverty line of US$1 a day for Indonesia and the Philippines, and US$2 a day for Malaysia and Thailand. With no change in distribution, poverty would double in Indonesia, and increase by 35 to 50 percent for the Philippines, Thailand, and Malaysia. Changes in inequality have the potential for an additional impact on poverty. A 10 percent rise or fall in the Gini coefficient—a significant change by past standards—in the distribution of income illustrates the possible orders of magnitude, and would have a major influence on projected poverty outcomes (see figure 5.3). In Indonesia, for example, a 10 percent worsening in inequality would cause poverty incidence to almost triple—from less than 7 percent (estimated) for 1997 to almost 20 percent in 2000. However, if inequality improved by 10 percent, poverty would remain largely unchanged.

A further scenario illustrates the impact on poverty of the most likely growth scenarios for each country for the period between 1998 and 2000 (based on consensus forecasts) with no change in inequality. Country-specific annual growth rates are used for this exercise to calculate poverty incidence in the year 2000. For Indonesia, the large expected declines in aggregate growth will force an additional 10 percent of the population into poverty by 2000. Thailand is also expected to be hit hard with a 20 percent increase in the headcount index. Malaysia and the Philippines are projected to show milder effects although simulations point to a deceleration of gains in the Philippines, and a reversal of a long-term trend of poverty alleviation in Malaysia. Any worsening in the distribution would aggravate poverty outcomes.

The scenarios show the powerful influence of overall economy-wide performance on poverty through its impact on both average incomes and inequality. But what can be said about actual short-term distributional outcomes?

It is difficult to draw general lessons from past episodes. But, it is possible to explore the relationships between short-term sectoral growth expectations and the structure of poverty. This has been done for Indonesia by applying projected sectoral growth rates to the structure of household incomes available from the consumption survey.[18] The simulations assume major declines in construction (-35 percent), commerce (-18 percent) and financial services (-18 percent), and a drought-induced slowdown in agriculture (1 percent), yielding a 12 percent decline in aggregate GDP. They examine only *impact* effects, and do not take account of either household responses or second-round effects, notably in the labor market.

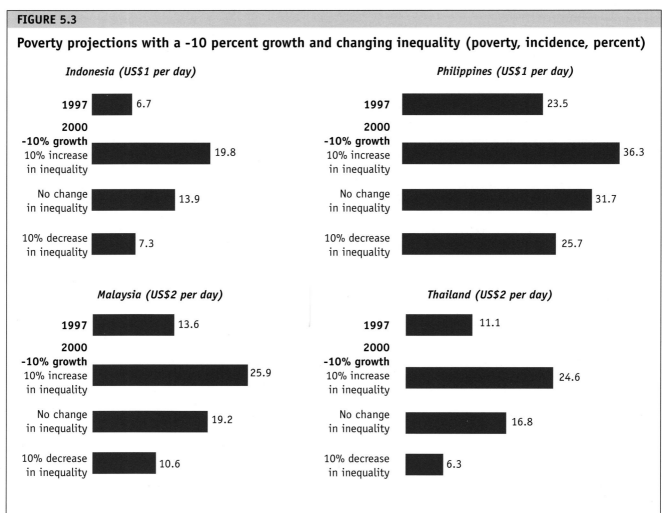

FIGURE 5.3

Poverty projections with a -10 percent growth and changing inequality (poverty, incidence, percent)

Indonesia (US$1 per day)

1997	6.7
2000 **-10% growth** 10% increase in inequality	19.8
No change in inequality	13.9
10% decrease in inequality	7.3

Philippines (US$1 per day)

1997	23.5
2000 **-10% growth** 10% increase in inequality	36.3
No change in inequality	31.7
10% decrease in inequality	25.7

Malaysia (US$2 per day)

1997	13.6
2000 **-10% growth** 10% increase in inequality	25.9
No change in inequality	19.2
10% decrease in inequality	10.6

Thailand (US$2 per day)

1997	11.1
2000 **-10% growth** 10% increase in inequality	24.6
No change in inequality	16.8
10% decrease in inequality	6.3

Note: A 10 percent contraction in aggregate GDP between 1998 and 2000 would add millions to the ranks of the poor. Beyond growth, distributional changes also strongly influence poverty outcomes. But, even with no change in inequality, the most likely growth scenarios (19880–2000) imply a halt or a sharp reversal in a 20-year long trend in poverty alleviation.
Source: Ravallion and Chen 1998.

TABLE 5.5
Impact effects on poverty in Indonesia
(percentage of population)

	Baseline January 1996	Forecast 12 percent contraction March 1999
Rural	15.0	17.6
Urban	5.0	8.3
Total	11.3	14.1

In Indonesia, a 12 percent decline in aggregate GDP in 1998 is estimated to increase poverty from 11 percent in 1996 to 14 percent in 1999, affecting both rural and urban residents. Even if second-round effects are ignored, most poverty in Indonesia would remain rural. Yet the immediate effect of the crisis is expected to be more acute in urban areas, resulting in a significant scale of urban poverty—8 percent in 1999—for the first time in many years (see table 5.5). Some groups would be particularly hard hit: the construction industry collapse would push poverty incidence among workers' families from 8 percent to nearly 31 percent; however, the actual number of people involved is relatively small. Of course, workers and households will respond and move into other labor markets. Java in particular has a high degree of inter-connectedness between rural and urban areas. If agricultural incomes remain stable, as the scenarios now indicate, there could be a rise in rural inequality: those with land would gain in relative terms, as those without land suffer from the general collapse in labor demand.

What is the cost of eliminating the crisis-induced increase in poverty through public transfers? Indonesia is used as an example in order to illustrate magnitudes.[19] In Indonesia, the increase in the poverty gap between 1996 and 1988–89 is estimated to be about 1.2 trillion rupiahs, or about 1 percent of 1988–89 budgetary revenues (see table 5.6). Restoring the poor's consumption to its 1996 levels would require a transfer of this amount. There are many ways this could be done, ranging from a universal (that is, untargeted) cash transfer to various targeted schemes, including some commodity subsidies and employment programs. An untargeted cash transfer, assuming no administrative expenses, would cost about 8 trillion rupiahs; the increased costs reflect the substantial leakages to the non-poor. But is possible to do better. Analysis done for Indonesia puts the cost of transferring US$1 through public employment (*padat karya*) programs at US$3.8–US$5.3; transfers through such programs would thus require 4.6–6.5 billion rupiahs or 3.5–5 percent of budgetary revenues. These are small amounts (as they usually are), especially when compared with the costs of financial sector restructuring. But administrative difficulties can be daunting.

TABLE 5.6
Cost of selected transfers to eliminate crisis-induced poverty in Indonesia
(in current rupiahs)

Increase in poverty gap between 1996 and 1998–99	1.2 trillion
Cost of closing the gap through:	
Universal cash transfers[a]	8 trillion
padat karya-type programs	4.6–6.5 trillion

Note: a. Assuming no administrative costs.

What can be done?

There is an urgent need to reduce crisis-induced welfare declines. Much can be learned from successes—and failures—in responding to past crises, notably in Latin America (see box 5.4). Public action plays an important role in six areas:
- The design of economy-wide policies

- The policy and institutional framework for labor markets and income security
- The level and pattern of economic and social services
- The interrelationship between corruption, institutions, and the social fabric
- The central role of public information.

How economic policies affect the poor

All the East Asian crisis countries are experiencing sharp drops in domestic demand. This is especially true of investment which is expected to fall by 10 percentage points of GDP per year, or more. This raises two questions: will future poverty reduction be jeopardized by the loss of productive investment and would compensating expansionary policies help the poor?

Protect investments that help the poor. An "investment pause," is a rational response to a shock. In times of significant drops in incomes, smoothing consumption over time, allowing more consumption now, at the cost of less in the future makes sense. This is of particular value to the poor, or those at risk of large drops in income. Indeed, a pause in investment helps explain why the contractions in Indonesia and Malaysia in 1984–87 occurred without any rise in consumption-based poverty.[20] However, there is a cost in lost future growth in incomes—high levels of investment were at the center of the growth process that brought such extraordinary poverty reduction in the past. While some investments, such as property development, were unproductive, on the whole East Asian countries put their investments to good use.[21]

Different investment categories have different effects on the poor. To reduce poverty it is particularly important to invest in building the capacity of poor children that, if foregone, may be irreversible, economic infrastructure in rural and periurban areas, and private investment in labor-intensive activities. The poor have a strong interest in overall policies that provide the confidence and financial sector preconditions for investment, especially in labor-intensive manufacturing and services. They also benefit from the protection of public spending for economic and social services that reach them, either directly as beneficiaries, or via the effects on labor demand during the construction phase.

What can East Asia learn from the experience of Latin America?

Latin America has experienced two major crises in the last 15 years; both began in Mexico and both involved substantial social costs. In the 1980s, Mexico's debt crisis spread throughout Latin America on the back of trade shocks and weak public finances. In 1995, Mexico's liquidity crisis, with its roots in excessive private lending, spread only to Argentina.

Latin America's two crises hit households hard. In the 1980s, real wages in Argentina and Mexico fell by nearly 40 percent, while poverty increased by more than 30 percent. In Chile, real wages shrank by about 15 percent and unemployment rose by 9 percentage points within a year. In the 1995–96 crisis, real wages in Mexico fell by more than 30 percent. In Argentina, unemployment rose by 6 percentage points and remained at around 18 percent of the work force for more than two years. As a result, poverty increased by more than 50 percent. Latin American governments responded with a variety of programs, but with the partial exception of Chile, the overall response was too little, too late. What are the lessons?

Solving social problems means putting them at the top of the agenda. Too often in the Latin American crises, policymakers' energy was devoted to restoring macroeconomic stability and implementing structural reform. East Asia has the opportunity to avoid this mistake by putting social issues at the forefront. Drops in income, employment, and public services have widespread and complex social consequences. Therefore, it is important to take action on a wide range of fronts, and make every effort to anticipate these consequences.

Targeting is crucial. Special funds set up in Latin America, designed to cushion the social costs of economic adjustment measures, had mixed success, largely due to poor targeting. Chile's large public works program was successful, in part because it set wages low to ensure that the most needy were the main participants.

Cutting subsidies can fuel resistance to needed reforms, especially of goods that are essential items to the poor, unless effectively targeted alternatives are offered. This was the case in the Mexican crisis.

Maintain support for core education and health services, or risk irreversible losses in human capital investment.

Social funds may be useful alternatives to government safety net programs, but only when the government really lacks the cohesion or resources to make programs work. Poor targeting, lack of genuine community participation, and long start-up times plagued many of the funds established in Latin America. There is now more knowledge on how to successfully design such funds well. But if established governmental or nongovernmental institutions work reasonably well, as they did in the past in most of East Asia, it is often better to work within that system.

More democratic systems do not hinder reforms. The Latin American debt crisis occurred as democratic rule was being restored in many countries and there were fears that democracy would either hinder painful but necessary reforms, or that the social disruption and violence associated with the crisis would hasten the return of authoritarian rule. Both fears proved groundless. Although the adjustment process was slowed in some of the countries by granting an independent voice to unions, legislatures, and other actors, the policy measures that were finally undertaken worked precisely because they had been endorsed by a more open civil society.

Source: International Herald Tribune, 5/29/98, "East Asia Can Learn From Latin America's Travails," by Nora Lustig and Michael Walton.

More expansionary macroeconomic policy can help the poor, but there are tradeoffs. The desire to reduce poverty provides a strong rationale to err on the side of expansion. Deciding what scale of fiscal expansion is desirable nevertheless, requires the assessment of other intertemporal tradeoffs since a more expansionary position could reduce investor confidence, or risk higher inflation and lower future growth prospects. Moderating the economic contraction is important for the poor, and may be distributionally favorable.[22] Equally important is how the fiscal expansion is financed. Deficit financing would be particularly damaging as earlier evidence from the Philippines and Brazil suggests that the poor are the hardest hit by high inflation.[23]

Interest rate policies will have mixed effects on the poor. High interest rates can moderate inflation, restore capital inflows, and encourage more labor-intensive new investment, all potentially pro-poor. But in East Asia in 1998, the short-term consequences of high interest rates on output, and consequently on labor demand, are central issues to the welfare of the poor.

Real exchange rate depreciations may help the poor in the long term, but short-term measures are important to cushion the effects of price increases. In an orthodox analysis, real exchange rate devaluation helps the poor because it favors agriculture—the most severe poverty is rural—and following a period of adjustment it encourages more labor-intensive investment choices. Evidence from the Philippines from the early 1980s, found that overall real exchange rate depreciation tended to help the poor. However, the effects of a depreciation are hard to generalize, and price effects are likely to be highly diverse across different groups in the population. Some of the poor are likely to benefit, such as rice farmers in Thailand, small holder export crop producers in the Philippines and Indonesia. But many poor households and the large informal sector are net consumers of tradable goods and would lose out due to a depreciation. The poorest groups in both Indonesia and the Philippines are rural wage earners. Where drought has hit and surpluses declined, many more farmers are net-consumers of food than in normal years. All of these groups are vulnerable to the price shock: the next section discusses the role of measures on the pricing and labor demand side to offset these effects.

Financial sector bailouts should not spare equity holders. Financial sector policy indeed matters to the poor; particularly in terms of the management of bailouts and effects it will have on microcredit availability. Dealing efficiently with a weakened financial system will require substantial infusions of public funds—perhaps on the order of 30 percent of GDP. While this is manageable within prudent debt limits, owing to low initial levels of public debt, it diverts vast resources from alternative uses to finance unsound or even corrupt past lending. The money behind the loans has mostly disappeared—whether to real estate follies or foreign bank accounts—and while it is standard practice to protect small depositors who are mainly middle income, few among the poor have significant levels of deposits. However, protecting larger depositors to avoid a collapse in confidence is costly. At a minimum, distributional and political economy considerations demand that owners of equity in the financial sector assume the losses before taxpayers. This is also more efficient since it reduces incentives for owners to go for high-risk lending strategies in the future.

East Asian countries have a wide range of financial institutions operating in most villages. Indonesia's BRI Kupedes is one of the world's largest rural micro-savings and credit programs. The Philippines has a wide array of small-scale programs, and there are significant, if smaller, sectors in Malaysia and Thailand. While many of these cater to better-off rural dwellers, they are sometimes an integral part of the village economy. There have been concerns over the consequences of the crisis for the micro-finance movement. A recent survey of micro-finance institutions found that in many, deposits continued to rise, possibly because those institutions were sound and were benefiting from a shift by rural savers out of smaller rural banks.[24] While some institutions enjoyed sustained high repayment rates, others suffered sharp rises in defaults. Efforts should be made to avoid adverse effects of selective deposit guarantees on rural financial services. While short-term measures may be needed to avoid the collapse of micro-finance institutions, sustainability should remain an important medium-term objective.

Invest in institutional transparency and accountability and in the skills of poor people to increase benefits from globalization and temper inequality in the longer term. The crisis has brought distributional struggles into the open. Three areas are likely to be important to tackle longer-term distributional concerns. First, a major societal concern is unfair or corrupt gains made by the wealthy and connected. Fostering sound corporate governance structures, open and accountable public sector institutions, and competitive procurement and bidding practices will all help, but this is a domain where the issues are often deeply linked to political structures and not amenable to technocratic solutions. Second, there is some evidence that the forces of global integration and technological advance target the skilled, leading to widening wage differences between those with and without education.[25] The appropriate response is not to withdraw from international engagement—countries that do this hurt themselves and their poor in particular—but rather to support inclusive and high quality educational systems. Third, in the past, growth largely bypassed populations living in poor areas. Policies that augment not only human capital, but also community capital in these areas, may be helpful, but in some cases outmigration may be the only effective strategy—as in the series of projects in inner China being supported by the World Bank.[26]

Food, employment, and income security

The economic crisis threatens the livelihood of the poor. When growth resumes, income gains and employment opportunities will expand, but this may take several years and there may be other temporary pauses in future growth. In the short term, the poor must be protected against drastic declines in consumption. Policy measures should aim at ensuring food security and preserving the purchasing power of vulnerable households. In the long-term, households will need assistance attaining income security during old age, and against health and employment shocks. There is also a need to reform labor market policies.

Ensuring food security

Food prices have increased sharply in most countries and there is evidence of food shortages in parts of Indonesia. Price increases on rice primarily affect the poor. In Indonesia and the Philippines, rice accounts for nearly 20 percent of household expenditures for the poorest one-fourth of households. In Indonesia, it jumps to 30 percent of household expenditures for the poorest decile. A relative increase in rice prices could hurt some of the long- and short-term poor and the non-poor hit by drops in labor demand; but it would protect households and villages that remain in food surplus.

Managing rice supplies to smooth price adjustments and avoid major price hikes may be desirable in the short-term—especially in light of the substantial overshooting of the exchange rate. This could moderate short-term inflationary and political shocks, and would help smooth consumption of poor, food-deficit households in rural and urban areas. But it is costly. In Indonesia, the cost of transferring one dollar to the poorest 15 percent of households through the rice subsidy is estimated at US$8.20—no better than a universal cash transfer.[27] If subsidies were limited to low-quality foods, the poor would benefit and rice transfer costs would be reduced to US$3.60 in Indonesia. Beyond the short term, there is a case for exploring additional mechanisms for targeting food subsidies—for example, through food stamps. While shifting to targeted subsidies is good policy in general, the Latin American experience shows that it is best not done during times of crisis.

Policies to keep food affordable are only useful if food is readily available. Ensuring that food markets function is fundamental to managing the effects of the drought. But, where markets are breaking down—as in parts of Indonesia—there is no alternative to direct food distribution. A combination of direct delivery to villages, NGOs, or religious organizations, food-for-work programs, and vulnerable group feeding may be desirable.

Sustaining the purchasing power of vulnerable households

Poor households are losing purchasing power due to falling labor demand and rising relative prices of some commodities. The case for rice subsidies was discussed above. Another major commodity, kerosene, is the most relevant subsidized fuel for the poor and near-poor in most of the region. Evidence from 1990 data for Indonesia indicates that kerosene subsidies do not effectively target the poor, with an incidence roughly in line with overall income distribution. Some phasing in of price increases may be justified on political grounds, but they should not compete with programs targeting the poor.

There are two possible approaches to addressing employment and income losses due to falling labor demand: stimulating demand for labor through labor-intensive public works programs, or providing income support for the unemployed. The question is how to target scarce resources and reduce leakages due to poor administration, graft, and access by non-targeted groups.[28] A combination of targeting techniques will need to be used, including geographic, household-based, and self-targeting.

Promoting public works. If designed properly, employment programs can help sustain the purchasing power of households in the short-term, and address long-term issues of declines in seasonal labor demand, as seen in parts of Indonesia. Indonesia is already expanding *padat karya* schemes and new programs are being launched in Thailand; there is also a long tradition of public employment in the Philippines.

Effective targeting depends on the wage. To ensure that the neediest choose to participate and to generate information on where demand for low-wage work is greatest, local market or lower wages should be used.[29] In Indonesia, for example, ongoing public works programs use a standard Rp7500 wage (in early 1998) that is often above local market wages, especially in depressed labor markets. Under these conditions, there is a greater risk that poor households will be rationed out of the programs. Finally, it is not clear whether women have full access to manual work in East Asian public works schemes. This is in contrast to many South Asian schemes, which often have high female participation. These concerns emphasize the need to include an overall monitoring and evaluation system and to conduct a continuing assessment of short-term transfer and targeting objectives.

Public employment programs have dual objectives—to transfer income and improve infrastructure. Chile in the early 1980s, and Korea in the late 1980s, are examples of public works programs that were allowed to fade away as demand fell.[30] Indonesia also has a history of using food and cash for work during periods of adverse shocks, then phasing them out. Experience from these and other programs provide insight into relevant design issues:

- *Balance urban and rural programs.* Both programs are effective but an urban bias misses the majority of the poor and may also increase commuter migration, or decrease return migration to rural areas, slowing efficient responses to the shock.
- *Balance cash transfer and productivity.* In the short term, cash transfers achieve the primary objective—to increase security. But over time, programs may become more focused on infrastructure production or maintenance.
- *Balance between male and female beneficiaries.* Programs should be open to participation by women but special efforts may be needed to publicize public works programs, review contracting procedures, and monitor the number of jobs given to them. In addition, certain design features—for example, piece rate payments instead of daily rates so women can work in groups and organize childcare—can encourage greater participation by women.
- *Workfare for the non-poor does not alleviate poverty.* Workfare programs for laid-off factory workers or young work force entrants with secondary or higher education who may choose not to take traditional manual work are counter to the rationale of workfare programs and do not alleviate poverty.

Some individuals and households, especially the disabled and households without adult labor, will fall through the safety net of public employment programs. In the short term, existing community channels, including religious organizations or NGOs, are probably the best source of support for these groups.

Unemployment benefits. With the exception of Korea, countries in the region do not have unemployment insurance schemes which could help sustain individuals temporarily out of work. But this is not the appropriate time to introduce insurance-type unemployment benefits schemes in the crisis-affected countries: benefits cannot be awarded immediately as insurance schemes generally require minimum periods of prior contribution and a waiting period before benefits can be paid. Furthermore, the introduction of payroll taxes, individual contributions, or both would increase the cost of labor (or the reservation wage) and hence depress labor demand (supply). New entrants to the labor market who are unemployed would not qualify for benefits. Finally, the scheme would largely benefit formal sector employees. Such schemes can be introduced upon a return to more stable economic conditions, but then attention should be paid to design features that reduce labor market distortions (see box 5.5).

In the short-term, a system of unemployment assistance benefits may be more appropriate. It would aim at alleviating poverty among the unemployed by providing, for a limited time, limited income transfers in the form of flat benefits at or near the poverty line. This scheme would have a better distributional impact than an insurance scheme and would also improve efficiency, as the low benefit level would motivate an early return to the labor market and have little impact on the reservation wage. Initially, this benefit could be provided to the unemployed without any means-testing. Later this feature could be adjusted or supplemented by means-tested family, health, and educational allowances. Unemployment assistance benefits can begin to be disbursed as soon as they are introduced and would be less demanding to administer than an insurance scheme. Nonetheless, targeting is a concern, especially if informal sector workers are to be included, and decentralized mechanisms for administration, including the use of NGOs, will need to be explored.

Training programs also provide some income support during the period of professional reorientation and skill acquisition, even though this is not central to their objectives. However, this is a costly means of transferring incomes with unclear distributional effects.

Reforming labor market policies

The question of international competitiveness may seem moot after the large exchange rate depreciations. But these exchange rate movements have been driven by capital flows and are likely to reverse once confidence is restored. The more fundamental issue is productivity growth which is linked to the work force skill level. In order to bolster skills levels, the education system must support increasing access to secondary education, but labor market policies are also important:

- *Encourage private job placement services to supplement inadequate capacity in the public sector.* There has been rapid expansion of public employment offices, from 53 in 1997 to 113 in 1998, with plans to reach 162 in 1999. This risks undermining the role of private employment services. In 1997, private employment services placed 1.8 million workers compared to with 0.3 million by public employment offices.
- *Exercise caution in launching or expanding active labor market programs.* International experience shows that these programs have little effect on employment probabilities or on future wage profiles. Some programs may be effective for particular categories of workers. Programs should undergo rigorous monitoring and evaluation and respond to feedback on effectiveness.
- *Change the government's role in vocational education and training.* The private sector can also provide both pre- and in-service training. Increased competition among training providers should increase quality, and the government can ensure regulatory oversight and increased choice for consumers—through vouchers, for example.

One of the most controversial issues in the labor market is the role of unions. This is because most East Asian countries have restrictive policies toward them, and there is considerable government involvement in industrial–labor relations. Korea was particularly repressive in the pre-democracy period, and reaped a bitter harvest in antagonistic labor relations. This experience carries lessons for other countries in the region.

A key challenge for most East Asian societies is designing a regulatory and institutional framework that provides strong rights for freedom of association and avoids granting entitlements that can cause resource

inefficiencies or crippling inflexibilities. This framework is likely to result in a more neutral role for governments—essentially setting the rules of the game. It should also encourage the primary locus of union activity to be in the enterprise in which collective action is most likely to help increase productivity and less likely to command monopoly wage increases.[31] There is a role for union confederations in situations where information or dialogue is necessary at a sectoral or national level. Developing effective mechanisms for industrial relations is complex; therefore, workers and employers must collaborate, often with state support, to design a sound and neutral regulatory framework.

Managing household insecurity in the longer-term

As the financial crisis demonstrates, temporary downturns in economic activity can lower living standards for most households. Unemployment, disability, and old age contribute to poverty in both industrial and developing countries, and worsen poverty among people who are already poor.[32] In most societies, coping with insecurity involves some combination of private savings, informal support, and employer obligations. Governments step in when these prove insufficient. Households may find it difficult to borrow to cover temporary drops in income; community support is ineffective when there is an economy-wide shock; and private markets for unemployment, disability insurance, and old-age pensions are often limited or absent.

East Asian countries have relied on informal income security systems for most households, especially the poorest. Private transfers help reduce inequality by providing old-age support, and alleviating the effects of disability, illness, and unemployment. Increasing urbanization and the growing importance of formal employment have eroded the informal support mechanisms. In addition, there is rising demand for social insurance to deal with household insecurity. A rapidly aging population further strains family-based support. These trends, plus rising incomes, suggest a much greater role for formal insurance in the future.

What kind of model should East Asian society follow to balance caring and competitiveness? The welfare state, created in Europe, is under attack because generous benefits and high taxes are associated with lack of competitiveness, slow employment growth, weak incentives for work, and high unemployment. In European and Central Asian transition economies, universal cradle-to-grave security is waning, while in China, the system for state workers is in need of urgent reform. In many South Asian and Latin American developing countries, the model of employment protection for formal workers is also under attack because job security regulations appear to protect insiders at the expense of comprehensive unemployment insurance, which encourages the growth of informal employment.

Countries in the East Asia region have an opportunity to develop schemes that avoid the labor market rigidities, inequalities, and fiscal problems associated with some models of social security, and they can learn from the abundant array of reforms being undertaken worldwide. The central principle is to minimize adverse labor market and fiscal effects by linking contributions to benefits. Most old-age pensions, certain health risks, and short-term unemployment can be covered under publicly-mandated insurance and savings schemes that embody this design feature; however, matching contributions directly to benefits is not always feasible or desirable. For example, most societies choose to provide for the poorest people far in excess of their potential to contribute—so the protection schemes incorporate an element of income redistribution. Socialized public action could be considered for core pensions for the long-term poor, to cover catastrophic health risks, and to provide social assistance for the destitute.

Maintaining economic and social services for the poor

During the crisis, the poor stand to suffer the most—especially from irreversible losses in potential education and health that will impede their participation in future recovery. Efforts to maintain purchasing power will help, but additional measures are needed to focus on keeping schools and health care affordable for poor households and quality of services intact.

Education. Studies find that public spending on primary schools benefits the poor. Beyond the primary level, incidence depends on the coverage of the education system. In low- and middle-income countries, spending on junior secondary schools is often distributed roughly in line with income. Spending on senior secondary and tertiary education tends to be unequally distributed, and can be even more unequally distributed than income, as is the case in Indonesia.[33] However, marginal changes in spending for junior secondary schools—and in richer societies, senior secondary schools—are likely to benefit the poor. Moreover, analysis has found that in the late 1980s, cuts in education spending were associated with significant drops in secondary enrollment. However, primary enrollment remained virtually universal. Since the current economic crisis looks much worse than the late-1980s slowdown, there is a strong chance that children in the region will be pulled out of school because of the actual and opportunity costs of schooling. Focus group results from Indonesia and the Philippines show that this is already happening. Also, there is a risk that for some children—especially of poorer households—this shift will be permanent. These factors suggest the need to:
- Preserve real spending on primary schools, and seek to maintain non-salary spending. In past episodes of

adjustment in other countries nonsalary spending has been the most vulnerable to cuts during a fiscal squeeze, with potentially high costs in quality.

- Increase targeted subsidies to encourage students to stay in secondary school, linked as closely as possible to income level. Subsidies could be structured as scholarships for the poor—perhaps using a village level mechanism for determining poverty—backed by broader loan-programs to finance fees for the non-poor (see box 5.6).

Beyond the crisis, the education system will shape the region's future work force and the competitiveness of its economies. Sustaining high quality and broad-based educational expansion is central to equipping workers with the skills for high productivity manufacturing and service industries, and to train them over the course of a working life.

As noted above, Korea has done exceptionally well in this respect, although recently it has been reassessing its education strategy with the goal of developing more creative and flexible skills. The pressure points in the region probably lie elsewhere, for example, in the relative neglect of secondary education in Thailand, upper secondary and college education in China, and in poor quality education in Indonesia and the Philippines. These issues are only partly a question of government spending priorities, as illustrated by Korea where secondary, and especially tertiary education is mostly privately financed. Institutional and policy reforms are required to foster the high quality schooling which includes the skills that will propel East Asian countries into the knowledge economy of the next century.

High quality schooling requires reforming curricula at the primary and secondary levels to emphasize team building, flexibility, and adaptability which are built on a foundation of literacy and communication skills, plus numeracy and analytical skills. To move their economies forward in the early 21st century, East Asia's young people will have to master the multitude of worldwide sources of information and be able to aggregate this knowledge to analyze and solve local economic and social challenges. Thailand and Malaysia are working to reform their education system to provide those skills, and China is also moving in that direction.

Health care. Analysis from Indonesia and Malaysia indicates that spending on health centers, particularly sub-centers, benefits the poor, but spending on hospitals

is unequally distributed. There are concerns that price hikes on imported drugs will lead to postponing or curtailing drug use, including vaccinations and HIV/AIDS drugs, and that sharp cost increases in private medical services are inducing greater demand for public services (see box 5.7). This suggests three measures:

- Preserve real spending on public goods or health activities with high externalities, such as vaccinations and vector control
- Maintain spending for health centers and sub-centers, especially for non-salary items
- Provide temporary subsidies for essential drugs, during a transitional period of exchange rate disequilibrium. Such subsidies will likely be weakly targeted to the poor, but they may still be justified in terms of protecting overall human resources.

Institutions, corruption, and the social fabric

East Asia's reputation as a model of reasonably efficient institutions and social stability has been shattered by the crisis. These important questions have arisen as a result:

- Are public sector institutions so riddled with corruption and cronyism that they can no longer deliver results?
- Is there potential for irreversible social breakdown including rising ethnic or other factional violence, and destruction of community and family behavioral norms?

Promoting effective institutions. Corruption and poor institutional performance shoulder much of the blame for the crisis. Corruption is a long-standing feature of most East Asian societies but its profile is on the rise with increasing public attention to international corruption rankings, and high-profile scandals from Japan to Vietnam. In the decades of extraordinary growth, corruption coexisted with reasonably effective institutions, from core macroeconomics management bodies to schooling services. Now most observers are concerned that public institutions are largely ineffective and driven more by private gain than the public good, especially in Indonesia.

Institutional capacity to deliver resources or services effectively is linked to broader governance concerns.

BOX 5.6

Preserving the poor's human capital during economic crisis: Indonesia's "back-to-school" campaign

Impact of the crisis. In April 1998, focus group discussions and school visits already indicated that poor schools and children were feeling the impact of the crisis. Reduced public funding for education, higher prices of schooling, and lower family incomes are expected to lead to declines in primary and junior secondary school enrollments among the poor. Estimating the impact of the economic crisis on enrollment is difficult as the crisis is unprecedented in terms of magnitude and depth. Econometric techniques have yielded relatively low impact effects ranging from an additional 115,000 to 260,000 7–12 year olds, and 173,000 to 270,000 13–15 year old children dropping out over time as a result of a 10 percent fall in per capita income. Estimates from the Government of Indonesia (GOI) point to much larger effects—an additional 890,000 and 640,000 children dropping out of primary and junior secondary schools, respectively, in just one year.

Whatever the precise figures, there is general agreement that the impact of the crisis on poor children will be severe. The strongest evidence comes from the much smaller economic shock of 1986–87 when education expenditures dropped and there were no special efforts to keep children in school; gross enrollment rates fell from 62 percent to 52 percent at the junior secondary level and took almost a decade to recover. Virtually the entire decline came from poor households. Also, non-salary spending per pupil fell sharply from Rp 23,000 (US$18) to Rp 6,000 (US$3) in real terms.

The response. On July 20, the government launched a 5-year national program to provide scholarships for poor children in basic education and block grants to schools serving poor communities. A coalition of ministers was formed to support the program. The World Bank is leading a multi-donor effort, which includes the Asian Development Bank (ADB), UNICEF, and bilateral agencies—AusAID, and Asia Europe Meeting (ASEM)—to support the program. Total cost of the 5-year program is approximately US$382 million, with an ADB contribution of US$86 million, and the remainder from the World Bank.

Seventeen percent of the poorest students will receive a scholarship of Rp 240,000 (US$30) in voucher form at the beginning of the school year. This is intended to cover school costs such as notebooks, uniforms, transportation costs, and school fees. Nationally, 2.6 million junior secondary students will benefit (about 17 percent of enrollment). Forty percent of primary and junior secondary schools serving the poorest communities will receive grants of Rp 2 million (US$250) and Rp 4 million (US$500), respectively. A total of 82,000 primary and junior secondary schools will benefit from block grants each year. Schools can use the grants to purchase instructional materials and other teaching–learning supplies, undertake minor repairs, and support poorer students by waiving formal and informal school charges.

Mass media and social mobilization effort. A nationwide TV, radio, and print media campaign was launched to ensure that parents and communities are aware of the program, to emphasize the importance of remaining in school, and to facilitate transparency in the use of funds and selection of recipients.

Targeting and election. Scholarships and grants will be allocated according to the poverty incidence of each district. Given the limitation of the quantitative data, this information will be coupled with local knowledge and the participation of NGOs and other members of civil society in the selection at the local level. Recipients will be selected by committees at district, sub-district, and school levels, which consist of parents, NGOs, other members of civil society, and government representatives.

Ensuring funds reach recipients. In order to ensure scholarships and grants reach intended beneficiaries, the program includes the following features: (i) funds will go directly from the (local) bank to students–schools—no intermediaries; (ii) a mass media campaign at the village level will inform communities and parents of the program and procedures; (iii) an independent agency will carry out quarterly monitoring; (iv) NGOs–civil society members will monitor the program; and (v) the government, the World Bank, and the ADB will evaluate the impact of the program on school enrollments and transition through focused surveys and the use of SUSENAS (Survey Economi Nasional).

Source: World Bank staff.

Indonesia has a reputation for institutionalized corruption. Evidence of institutional weakness is scattered, but it ranges from the low quality of education, health, and other services to the widespread concerns over graft and the view that local resource allocations are determined by political power, rather than developmental and social needs. However, East Asia could not have enjoyed the massive advances in social conditions if government services—a major part of this effort—were useless. Careful micro studies are scarce, but reveal a mixed picture. A comparison of public irrigation workers in Korea and India found sharply better performance in Korea.[34] Recent research on local institutions in Indonesia finds reasonable performance of public institutions at the village level—although they are significantly worse than genuine community organizations.[35]

Institutional reform is a complex and long-term process and decentralization, though often desirable, is not a panacea, especially during the crisis when effective

The crisis and health: Common set of problems

Medical costs are increasing. Exchange rate depreciations have meant large increases in medical costs given the high import content of pharmaceuticals, including vaccines and contraceptives. In Indonesia, imports account for 60 percent or more of the pharmaceuticals used in the country, and drug prices have reportedly increased two- or three-fold. This change in relative prices is unlikely to be fully reversed, and will require long-term adjustments in drug consumption patterns.

Private consumption expenditure is falling, particularly among the rising numbers of unemployed. Many households are less able to pay for the out-of-pocket cost of medical care, whether provided by the private sector or by public sector facilities that typically charge user fees. This is important because private spending finances approximately 50 percent of aggregate health expenditures in East Asia. There is already evidence that private sector users are switching back to the subsidized public sector, while some potential users—especially among the poor—may have to switch to lower quality providers, or even forego medical care entirely.

Public health expenditures are declining. Budgetary pressures can reduce public subsidies, which protect the poor from the increased financial risks of illness. This either increases financial hardship, or reduces use of medical services. Moreover, increased demand for public services from former users of private facilities could divert public subsidies from the poor. In the long-term, cuts in operations and maintenance outlays will also undermine the productivity of the public infrastructure. Reduced public expenditure also threatens priority public health programs, such as immunization against childhood diseases and TB control. Indonesia's past experience with fiscal adjustment in the mid-1980s demonstrates the vulnerability of public health programs to public expenditure cuts.

Source: World Bank staff.

delivery is unusually urgent. In fact, decentralization of power to local elites can make things worse, in part because they generally have weaker technical resources at their disposal.[36]

Evidence from reforms in Latin America and elsewhere has emphasized both giving more choice and more voice to the communities who use services.[37] In particular, a diversified approach to delivery of transfers, involving government, civil society and religious institutions, can help reduce the risks of relying on only one delivery channel. Restrictions on local associations should also be lifted to encourage entry and competition. This can be complemented by measures to foster genuine participation of communities in the choice, design, implementation, and evaluation of projects. Social funds were much used in Latin America and Sub-Saharan Africa in response to adjustment. There is also increasing evidence that they perform best when there is genuine participation of local communities.[38] Finally, public and independent information, including monitoring by civil society organizations, is a potentially valuable source of increased accountability.

Responding to a deteriorating social fabric. Responses to the added pressure on social relations within the family and the community vary across different societies (see box 5.8). It is too early to assess the consequences of social changes, but there are parallels in other communities under economic crisis. Economic decline contributed to rising violence in urban Latin America in the 1980s. Even though much of the region recovered economically in the 1990s, violence remains high, with widespread economic and social costs. Studies in poor urban communities in Ecuador, Hungary, Zambia, and the Philippines found:[39]

- Increased work for women and children
- Increased pressures on women and older girls—mothers work more, so girls must care for younger children
- Increased substitution of private for public services—including in health and education
- Increased street violence, especially amongst young males
- Increased domestic violence, especially in households hit by falling employment or declining incomes.

The study found varied effects on informal community functioning. Moderate pressure could lead to heightened mutual support—increased use of social capital—whereas severe pressures were more likely to breakdown community-based coping mechanisms.

These substantial costs underline the importance of restoring the national and local economic environment crucial for social functioning. Specific responses could include:

- Identifying the vulnerable groups and focusing action to respond to their needs. For example, heightened pressures on women can be relieved by reducing the demands on their time, such as improving water supply or childcare facilities.

- Facilitating participatory processes mediated by government, NGOs, or religious institutions—both to set local priorities and to support informal networks.
- Supporting innovative action to reach groups at high risk. For example, concerted action by civil society is often the only way to reach children forced into exploitative work. In Brazil local groups use theater, music, and other forms of community engagement to reach kids on the streets, diverting them from gangs to more productive forms of social capital.

Increasing monitoring, diagnosis and public information

Assessing public actions will provide information which will be included in ongoing redesign of these programs. Public information will provide checks on transfers and strengthen an informed public debate on developments, effects of programs, and tradeoffs. *Programs that monitor overall welfare should rely on a mix of instruments.*

- *Conduct regular surveys of living conditions and vulnerability* covering wages, unemployment, relative prices, food prices, drought effects, social indicators, and nutritional levels. In many East Asian

countries, such short-term information is incomplete. The Philippines has a Social Weather Station system based largely on subjective assessments. Indonesia recently dropped the quarterly version of its labor force survey because there was little seasonal variation. Thailand still has a quarterly labor survey and Korea's labor survey now takes place monthly.

- *Conduct complementary assessments* of household and commentary conditions using participatory techniques which substantially enrich the understanding of coping strategies. These assessments should form part of the ongoing interaction between hypotheses and data, and could influence the design of survey questionnaires.
- *Use existing data sets* to match economy-wide trends to the structure of income and spending, and to analyze past household responses to changes in specific parameters, such as price elasticities of schooling costs. In most countries, existing data includes consumption or income surveys.

Monitoring and evaluation of public action are important short-term goals to ensure that intended effects reach targeted groups and to redesign programs. In the medium to long term, monitoring and evaluation help to bring poor regions and groups into the develop-

---

BOX 5.8

Erosion in social capital

Most citizens see the crisis eroding social capital—trust, reciprocity, and networks of support. Nevertheless, in some communities social cohesion may actually strengthen as poor communities discover resourceful ways to overcome their problems. For example, in Davao, on the island of Mindanao, in the Philippines the community initiated a savings scheme to cover the costs of festivals, and a self-policing program (ronda) was developed in response to increased crime.

Conflict. In all countries, NGOs identified increased conflict—within the household, the community and society at large. Increased pressure led to a rise in domestic conflict, and "loan sharks" in Bangkok resorted to violence against people who could not repay their debts. NGOs also expressed concern over the potential for social unrest, a concern which manifested in Indonesia where there has been extensive ethnic violence, including rape, against members of the Indonesian-Chinese population,.

Vulnerability and insecurity. In Teparak, a slum settlement in Khon Kaen, in northeastern Thailand, a focus group identified a

breakdown in community trust within the last six months. Increased competition for jobs meant that neighbors who once cooperated were now competing. Theft, violence, and other crimes were on the rise. Some children had been forced by their parents to drop out of school to guard their homes because both parents were now working outside and break-ins had increased.

Isolation. All focus groups reported a general feeling of uncertainty and isolation. Many said that although the poor had benefitted from improved social welfare, they still felt excluded and felt that they had not received their fair share of the economic growth. Many blamed the rich for the current crisis and were unable to understand why the poor should carry the burden. In Teparak, a community leader added, "The crisis has happened too quickly and has left us confused, puzzled, and let down. We have been laid off but given no explanation."

Source: Robb, Caroline (1988). "Social Impacts of the East Asia Crisis: Perceptions from Poor Communities." Prepared for East Asian Crisis Workshop, July 1998, IDS, University of Sussex, UK.

BOX 5.9

The World Bank's efforts to help the poor

The World Bank is helping governments re-initiate growth to manage the social consequences of the crisis; protect public expenditures targeted for the poor; enhance the quality of social services; improve design and financing of social funds; strengthen social security systems for the unemployed and the elderly; and address key institutional issues. The World Bank's most important activities include:

In **Thailand**, a US$300 million loan for a social investment project will fund job creation for the poor and the unemployed through existing labor-intensive government programs; expand training for the unemployed; support low-income health insurance schemes, small-scale community projects, and larger municipal projects; and set up a monitoring system to evaluate the impact of the crisis and of public action on the poor. The loan is expected to create roughly one million months of jobs and an equivalent amount of training. Also, a national poverty map will be drawn based on available statistical data and a nationwide systematic participatory assessment which will be an important input for the policy debate on safety net mechanisms

In **Indonesia,** the World Bank has restructured some of the existing portfolio to redirect savings to support income generation and meet basic needs (about US$320 million). A Structural Adjustment Loan (SAL) of US$1 billion includes a component to protect the poor through expanded labor-intensive public works programs; actions to ensure the continued availability of key goods with only modest price increases; and initiatives to maintain access to quality basic education and health. In particular, to ensure continued high enrollment rates for children through the first nine years of school, the government is to provide scholarship funds for 2.6 million needy junior secondary school students. A US$275 million poverty project for rural areas (the Kecamatan Development Project) has been approved, and a similar project is under preparation for the urban poor. Discussions are also underway regarding an agriculture sector adjustment loan to support reforms. The World Bank has also intensified analytical and participatory work on poverty to help underpin these two operations and will help finance a follow-up to the Indonesia Family Life Survey (IFLS). This will allow monitoring of the living standards of a sub-sample of households that were already surveyed in 1993 and 1997 and help assess household-level coping strategies in response to the crisis.

In **Korea**, the US$2 billion SAL approved in March 1998 includes an important program on labor markets and social safety nets. The program incorporates measures to increase flexibility in the labor market while extending coverage of unemployment insurance to employees in small-scale enterprises; improve poverty monitoring and protect poverty-related public expenditures; and reform the pension system. A conference in July 1998 focused on lessons of international experience in labor market policies. Another conference will focus on pension fund investment policies. A second SAL for US$2 billion will help deepen these reforms and start addressing issues in health financing and health care.

In the **Philippines**, three new loans (US$79 million) to increase the incomes of the poor and to provide basic services were approved in March 1998. Two other loans totaling US$130 million are scheduled for approval in the second half of 1998; these projects finance infrastructure development and will increase job opportunities and access to basic services for relatively poor local government units. The World Bank has carried out a rapid social assessment to gauge the effects of the crisis and understand household coping strategies. Poverty work scheduled for fiscal 1999 will have access to the results of the 1998 Income and Expenditure Survey (FIES is carried out every three years) and contribute to the early implementation of the Annual Poverty Incidence Survey. This will provide a useful analysis of the short-term impact of the crisis, evaluate the effectiveness of government policies to alleviate poverty, and provide policy directions for the future.

In Malaysia, a US$300 million Economic and Social Sector Loan approved in June 1998 will support a reduction in the fiscal surplus from 2.5 to 0.5 percent of GDP by increasing public expenditures for the social sectors. The loan seeks to protect budgetary spending for education, health, rural infrastructure, and to increase expenditures on social safety net programs aimed at providing direct support to the poor (free housing and food supplements) and income generation through small grants. Longer-term issues about the adequacy of formal safety nets and the governance structure of the Employee Provident Fund will be addressed through a Technical Assistance Loan and economic and sector work. In addition, a CEM is nearing completion—the first since 1993. It includes an overview of poverty and the social safety net in Malaysia, analyzes how the poor may be affected by the downturn, and recommends action to cushion the impact of the crisis on the poor.

In **China**, ongoing work in labor market adjustment focuses on policies needed to address the un(der)employment problem. While the problem is becoming acute largely due to accelerated reforms in the state-owned enterprise sector, the slowdown in aggregate demand, which will be exacerbated by the impact of the regional crisis, is also having an impact. A workshop will discuss the effectiveness of active and passive labor market policies in addressing employment problems.

In **Cambodia**, a study is underway to examine the impact of the regional crisis on Cambodia and **Lao People's Democratic Republic**, including an analysis of the social impact, in particular through rapid social assessments.

A **region-wide** initiative is being launched to analyze issues in pensions policy and administration. A November 1998 conference will bring together countries of the region to explore common issues and frame an agenda for future work. This would be followed by a conference in the spring of 1999 on governance of pension funds in East Asia.

Source: World Bank documents.

ment process. All programs can benefit from combined quantitative and participatory monitoring. Participatory monitoring is particularly valuable for increasing effectiveness by strengthening community involvement, and increasing efficiency by scrutinizing the use and allocation of funds. Public information strengthens accountability. For key programs, especially those with uncertain impacts, structured evaluation that uses samples of participating households and controls is also important for maximizing the benefits of scarce resources. For example, structured evaluation is likely to cover the targeting efficiency of public works, the incidence of new subsidies for education, and the effectiveness of geographic targeting to alleviate poverty.

Notes

1. This section draws on Ahuja and others. (1997).

2. World Bank (1993a), Birdsall and Sabot (1993); Teranishi in Aoki, and others. (1996).

3. World Bank (1995), *World Development Report*; World Bank (1996a), *Involving Workers in East Asia's Growth*.

4. See Kim and Topel (1995).

5. See Ranis (1995).

6. After controlling for incomes and other characteristics (though in Indonesia this may partly reflect unusually rapid income growth, and the slower response of mortality. See Filmer and Pritchett (1997): Indonesia is a outlier (that is, had high child mortality) after controlling for incomes; the Philippines and Korea were negative outliers after controlling for incomes and a set of other characteristics, including female education and inequality.

7. Vinod, and others. (1997), *Everyone's Miracle? Revisiting Poverty and Inequality in East Asia.*

8. Jalan and Ravallion (1998), "Spatial Poverty Traps?"

9. These comparisons ignore issues such as equivalence scales and differences in regional cost of living. In addition, the discussion includes Ginis for both expenditure and income distribution. Since income distributions are generally more unequal than expenditure distributions, comparisons are only valid across indices for the same concept (for example, over time). Similarly, we include distributions per household and per individual, which, once again, are not strictly comparable. The figures merely indicate trends.

10. Preliminary analysis of the 1997 survey yields an increase from 45.1 in 1994 to 49.6 in 1997 in the Gini coefficient for family income. The absolute numbers are not to be compared to the data cited in

table 5.4 which are Gini coefficients for expenditure per capita but the trends and magnitude of changes are likely to be similar.

11. It is possible to observe increases in inequality that leave poverty unchanged for a given level of average income. A mean-preserving spread originating from a transfer from an individual above the poverty line will increase inequality but not affect poverty.

12. Galenson (1992).

13. World Bank (1996a), *Involving Workers in East Asia's Growth*.

14. Based on The Social Weather Station surveys of Filipino's perceptions of their poverty.

15. Robb, Caroline M. "Social Impact of the East Asian Crisis: Perceptions from Poor Communities." Paper prepared for the East Asian Crisis Workshop, IDS, University of Sussex, UK, June 1998.

16. See World Bank (1990), *World Development Report 1990*.

17. Projections are from Ravallion and Chen (1998). The methodology involved updating household data to 1997using actual or estimated growth in average consumption or income per capita, but assuming no distributional changes since the most recent survey—survey years are reasonably recent: Indonesia (1996), Malaysia (1995), Philippines (1994), and Thailand (1992). Then a range of projections for the entire distribution was developed by assuming alternative values for growth in the mean level of consumption or income and the degree of inequality. The change in inequality was estimated in terms of the shifts in the parameters of the Lorenz distribution to produce a certain percentage change in the Gini coefficient.

18. This scenario was developed by Benu Bidani, as part of ongoing work on poverty in Indonesia. The poverty line used in this scenario is different from the international poverty line of US$1 per day at 1985 prices. For a full discussion see "The Poor in Indonesia's Crisis," mimeo, World Bank, 1998.

19. The Poor in Indonesia's Crisis, World Bank, August 1998.

20. See World Bank (1990), *World Development Report 1990*, Chapter 7.

21. See Sarel: that actually found rising investment efficiency in Indonesia in the past decade.

22. Work on the U.S. finds that unemployment disproportionately hurts the poor (Blinder and Blank). Similar effects may explain some of the recession-linked increases in inequality in Latin America.

23. Blejer and Guerrero (1991); Ferreira and Litchfield (1997).

24. Banking with the Poor Network Newsletter, Issue No. 11, June 1998, pp1–8. Foundation for Development Cooperation, Brisbane.

25. See Wood, Pissarides and Tan and papers in the WBER volume.

26. See Staff Appraisal Report for South West Poverty Reduction Project (May 1995) and Qinba Mountains Poverty Reduction Project (May 1997).

27. This is because per capita rice consumption does not vary much across expenditures classes (the bottom 15 percent of the population consumes 13 percent of the rice). Therefore the proportion of the rice subsidy that goes to the poor (and the leakage to the non-poor) is no different from that of a general cash transfer.

28. In some poorly targeted Indian programs 6–7 rupees are spent to transfer one rupee to a poor household. Radhakrishna and Subbarao (1997).

29. See Ravallion (1998), "Appraising Workfare Programs."

30. See Subbarao, and others, 1997.

31. Pencavel (1995) and World Bank (1996a), *Involving Workers in East Asia's Growth*.

32. World Bank (1996a), *Involving Workers in East Asia's Growth*.

33. See World Bank (1993), *Indonesia: Public Expenditures, Prices and the Poor*.

34. Robert Wade, 1994. "The Governance of Infrastructure: Organizational Issues in the Operation and Maintenance of Irrigation Canals." World Bank.

35. Preliminary work on the Local Institutions study from World Bank staff and the study team.

36. *World Development Report*, 1997.

37. Carol Graham, *Private Markets for Public Goods: Raising the Stakes in Economic Reform*, Brookings Institution Press, Washington D.C., 1998.

38. See Narayan and Ebbe, 1998.

39. "Confronting Crisis. A Summary of Household Responses to Poverty and Vulnerability in Four Poor Urban Communities," ESD Studies and Monograph Series No. 7, 1996.

chapter six

Environment in Crisis: A Step Back or a New Way Forward?

Bill Wong knew he was a pioneer when he opened a hi-tech plant in Malaysia's remote state of Sarawak, on Borne, early in 1996. He never guessed that 18 months later, his $115 million factory would be engulfed in smoke from some of the largest forest fires in history. Air pollution readings of over 500—on a scale on which readings above 301 are considered dangerous—kept half of Wong's 800 local employees off work. They weren't all needed anyway; thick smoke and haze closed the airport and port in nearby Kuching, making it impossible for Wong to get his products to customers. The plant cut back to a single shift, and the workers who did show up were given bonuses and free meals because food prices had jumped as much as 500 percent. Wong's problems are just a small fraction of the mounting costs Southeast Asia faces from the environmental disaster created by the epic forest fires in Indonesia. The full impact of the damage probably won't be known for another decade or more.—Murray Hiebert and others, "Fire in the Sky," *Far Eastern Economic Review*, October 9, 1997.*

In the years prior to the crisis, people in East Asia were beginning to see that a "grow now, clean up later" policy had resulted in high environmental costs. Many city dwellers were suffering from respiratory and related illnesses or dying prematurely as a consequence of poor air quality. The loss of forests destroyed natural habitats, contributed to soil erosion, and increased the severity of floods. Water pollution was threatening the productivity of irrigated agriculture and fisheries, increasing the costs of industrial enterprises, and endangering the population, especially infants and young children. Environmental problems grew worse in 1997 because of a severe drought and the severe forest fires in Indonesia. Without radical changes it seemed certain that environmental problems would only increase as a result of continued economic growth and urbanization.

This realization prompted efforts throughout the region to improve environmental management. The question now is will the financial crisis in East Asia cut short these efforts, or will it provide an opportunity to follow a better path in the future? This chapter highlights the choices that East Asian countries face and suggests ways in which past trends can be reversed without jeopardizing the prospects for economic recovery and future growth.

The immediate effects of the crisis have been beneficial for the environment. Sharp declines in incomes and industrial output have substantially reduced air and water pollution caused by vehicles and industry. World market prices for timber and many other natural resources have collapsed, reducing the profitability of current production and increasing the return that may be obtained by postponing production into the future. These short-run adjustments are consistent with what is known about the environmental impact of previous economic crises (for example, the Latin American debt crisis of the 1980s, the collapse of Eastern European and Central Asian socialism at the beginning of the 1990s, and the Mexican financial crisis of 1995).

However, many observers are concerned that a prolonged recession will increase pressures on natural resources. Fewer jobs and lower urban incomes will force marginal urban residents to move back to rural areas, which will accelerate the conversion of forest land to agriculture and increase the stress on critical resources, such as fish stocks and water resources. The consequences of poverty and desperation will be reinforced by shifts in relative prices if the crisis results in a large depreciation of exchange rates. A depreciation would increase the income obtained from exploitation of natural resources, such as forests and minerals, to pay debts or sustain consumption. Finally, public budgets for environmental management may be drastically reduced as concern for the environment is crowded out by other spending demands and expenditures for financial restructuring.

Policy makers must give priority to restoring growth. Environmental improvements resulting from an economic crisis and a slowdown of growth are realized only at a great cost. Restoring economic growth is essential in order to reverse the declines in incomes, reduce poverty, and establish a proper balance between people and the environment. However, this priority for growth should not necessarily be incompatible with plans to protect the environment.

Environmental dimensions of the crisis

The economic and financial crisis in East Asia has an environmental impact larger than similar crises have had elsewhere in the world because of the cumulative effects of serious natural resource mismanagement in the past and a severe drought in many parts of the region. The effects of the drought have been most serious in Indonesia, the most rural of the main economies in Southeast Asia. Here crop failures and the resulting acute distress in many rural communities have exacerbated the sharp decline in employment and income in urban areas. Furthermore, the drought, combined with mismanagement of forests in Kalimantan and Sumatra, led to extensive and prolonged forest fires. These fires have not only damaged forests but also spread air pollution over many areas of Southeast Asia, which is likely to have caused many deaths and even more cases of respiratory illnesses (see box 1).

Whether the economic and financial crisis was provoked or exacerbated by deep-seated flaws in the region's economic policies remains a matter of dispute. However, the combination of economic crisis, drought, and forest fires highlighted the weaknesses of existing institutions and policies that address the management of natural resources and environmental problems. The following examples illustrate this problem.

• Parts of Indonesia, especially eastern Java and the outer islands, have always been prone to drought. Irrigation systems, grain storage and distribution networks, employment programs, and other mechanisms have been developed to mitigate the effects of droughts or to assist those affected. However, overuse of water resources combined with reckless discharge of industrial and municipal wastewater mean

Forest fires: A symptom of deeper problems

Southeast Asia was hit by two blows in the summer of 1997. The rapidly spreading financial crisis was accompanied by an environmental crisis as forest fires burned out of control, razed more than 300,000 hectares of forest in Indonesia, and spread a thick pall of smoke over large parts of Southeast Asia. The smoke, combined with urban air pollution from traffic and other sources, caused immense health, social, and economic damage.

Estimates suggest that more than 7 million people were affected by the haze, which caused premature deaths and severe respiratory illnesses such as asthma and bronchitis, as well as other health problems, including sore eyes and skin rashes. There were also substantial economic costs, large timber losses in Kalimantan, and a decline in tourism by as much as 30 percent in Singapore and Malaysia, and even more in Indonesia. More than 1,100 flights had to be cancelled as airports were closed because of smog.

The fires were a symptom of deficiencies in forest management and policies as well as regulations about converting land to various commercial uses. The forest fires actually began as controlled fires, a widespread practice used by farmers to clear areas for cultivation. The severe drought caused by El Niño allowed the fires spread to secondary and primary forest areas, grasslands, and peat bogs.

The forest fires dramatically demonstrated the fundamental structural problems of the Asian economy. They forced political leaders to acknowledge the costs of current practices of commercial timber and plantation companies. Furthermore, they offered an opportunity to introduce reforms that addressed land use, forest conversion for plantations, clearing methods, and a renewed effort to enforce existing environmental restrictions.

Although the use of fire to clear agricultural land was outlawed in Indonesia in 1995, agricultural and forestry authorities lacked the political commitment and budgetary resources to implement the ban and effectively stop the practice or prosecute violators.

Source: World Bank staff.

that agricultural and industrial production has become more vulnerable to water shortages. Programs and policies to alleviate poverty have been disrupted by economic and fiscal stringency as well as a lack of institutional capacity.

- Mismanagement of water resources has become an increasingly serious problem in Thailand. In the main watershed areas, deforestation has changed seasonal patterns of run-off and irrigation systems have been expanded beyond their capacity into rivers to provide reliable supplies. Congested urban areas pollute the water supply needed by users downstream. In coastal areas, the loss of mangroves and the misuse of pesticides threaten the future of the once-thriving shrimp industry. Drought and economic shocks serve to reinforce these problems at a time when the government and the Thai people wish to rely more heavily on agriculture for income and exports.

- Many cities in Southeast Asia suffer from severe air pollution. The main sources of particulate emissions are diesel trucks and buses, two-stroke motorcycles, and the burning of kerosene for domestic use. Reducing air pollution requires a determined effort to penalize polluters, in addition to ensuring that fuel prices and other economic incentives correctly incorporate the social costs of pollution. In the short run, the crisis has resulted in a slight decline in traffic and pollution, a consequence of the decline in economic activity. However, long-term initiatives are stalled by a political reluctance to adjust fuel prices and insist on changes in behavior by vehicle operators, even though these changes could generate a significant fiscal savings in the medium to long term.

- There are widespread concerns that massive currency devaluations will accelerate the wholesale cutting of mature tropical forests in order to realize the immediate value of the timber and also to plant alternative crops with a better economic return. Weaknesses within the institutions responsible for forest resources combined with low fees and taxes on the exploitation of such natural resources previously have led to large-scale mismanagement. This problem may get worse as a result of short-term pressures for immediate revenues.

East Asia's financial crisis and environmental problems have similar roots: rapid growth without proper safeguards, policies, and controls (see figure 6.1). In the financial sector, the capacity of governing institutions and policies has been outpaced by the growth of capital flows and lending. In the environmental arena, growth has outstripped both the absorption capacity of the environment and the speed with which policies and institutions can respond to new challenges. Collusion ronmental policies and institutions, just as financial contagion has exposed the basic weakness of East Asian financial systems. Financial instability has had a direct and visibly large impact on economic activity and incomes. In comparison, the effects of environmental neglect are more insidious and will become evident over time, although they may be no less significant in the aggregate. Still, the short-term costs of the drought and the forest fires have been significant (see box 6.2).

The weaknesses of environmental regulatory arrangements in East Asia have been recognized for some time, even if progress in dealing with the problems is frustratingly slow. There is, however, another environmental dimension to the crisis that could pose a more serious obstacle to economic recovery and future growth.

All of the countries in Southeast Asia historically have relied heavily on exports of raw or processed natural resources to import capital goods and support economic growth. Rice, palm oil, timber, metals, oil, and gas have been and remain important or even dominant sources of foreign earnings. Agricultural growth has been supported by an expansion in the area of land under cultivation—from 15 percent to 23 percent of total land area in Malaysia from 1980 to 1994 or from 36 percent to 41 percent in Thailand over the same

between segments of government and parts of the private sector exerts pressure on agencies to provide subsidies, directed credit, and exemptions from regulations, compromising their ability to enforce appropriate standards of prudence and good performance.

As an example, East Asia's forest sectors are poorly developed as defined by the degree of integration with the national economies, levels of technical and economic efficiency, environmental performance, and quality of governance.[1] Asia has lost more of its forest area in the last 30 years than any other region in the world (see figure 6.2). Regional forestry policies fail to recognize the scarcity of forest resources and, hence, provide inappropriate incentives to preserve them. The low stumpage fees, which should reflect the rents earned by extracting timber, are the most serious problem because they encourage deforestation and lead to the impoverishment of forest-dependent communities. In Indonesia, the annual sustainable harvest is estimated at 22 million m^3, but the annual production of forest products is in excess of 40 million m^3. Furthermore, land management regulations, especially those related to the conversion of forests to other uses such as tree and plantation crops, are largely ineffective.

Thus, the combined effects of drought and economic crisis have highlighted pre-existing deficiencies in envi-

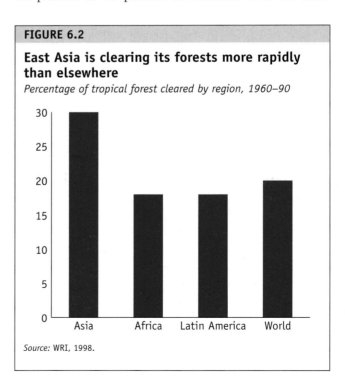

FIGURE 6.2

East Asia is clearing its forests more rapidly than elsewhere

Percentage of tropical forest cleared by region, 1960–90

Source: WRI, 1998.

The costs of drought and forest fires

Some attempts have been made to estimate the scale of the environmental damage caused by the combination of drought and forest fires during 1997–98. Many of the figures quoted are based on anecdotal evidence or on questionable economic assumptions. Thus, to date it is only possible to offer a partial assessment of these costs.

Loss of forest cover. There are widely differing estimates of the forest areas that were burned during 1997–98. The Indonesian government put the loss at under 300,000 hectares, although other estimates are 10 times this figure. The losses caused by the forest fires represent the difference between the value of the land burned as forest and its value as agricultural or plantation land, adjusting for any residual value of timber that can be recovered. The average value of forests is about US$1,500 per hectare, of which nontimber benefits account for up to 20 percent. Assuming an average loss of 50 percent of the timber value as well as all of the nontimber benefits, this yields an overall loss of US$270 million for a low estimate of the area burned or up to US$3 billion for a high estimate.

Loss of plantation and small-holder perennial crops. The forest fires are believed to have affected up to 1.7 million hectares of perennial crops, particularly in palm oil plantations and a variety of small-holder crops. The difference between the value of land with mature crops and in alternative agricultural uses is US$2,000 to US$3,000 per hectare. Not all of the crops have been lost; land that was burned can be replanted and will reach maturity in five to eight years, depending on the crop and the management

regime. Thus, the average loss is unlikely to exceed US$1,000 per hectare or US$1.7 billion in total.

Loss of annual crops. The drought and haze caused by forest fires have had a severe impact on agricultural production in large parts of Indonesia and the Philippines, with lesser effects in Thailand and Malaysia. Assuming that the maximum loss of agricultural value-added in Indonesia and Philippines was 5 percent with a 2 percent loss in Thailand and Malaysia, the total loss of value-added will have been about US$2.6 billion. Most of this loss was the result of the drought rather than the fires.

Health damages. Many people throughout Southeast Asia were exposed to high levels of particulates and other pollutants for periods of three to six months as a result of the haze caused by the forest fires. It is reported that peak levels of smoke exceeded 6,000 micrograms per cubic meter. Based on estimates of the damage caused by previous severe smogs and by current air pollution in China, the lowest reasonable estimate of the cost of this air pollution would be 2 percent of GNP for Indonesia, Malaysia, and Singapore and 1 percent for the Philippines. This amounts to US$6.7 billion, but the true cost may be two to three times this figure.

Overall, the cost of the damage caused by the combination of fire and drought may have reached US$12 billion to US$14 billion, with one-half of this cost relating to the health cost of air pollution caused by fires. This amounts to about 2.5 percent of total GNP for the main countries affected.

Source: World Bank staff.

period. It could be that the economic slowdown is linked to the diminishing scope for extensive growth and a failure to use resources more productively. If this is the case, economic recovery and growth may depend on a willingness to accept greater exploitation of natural resources, even if only on a temporary basis.

The evidence concerning the extent to which economic growth in East Asia has been financed by depleting natural capital is mixed (see box 6.3). Natural resource rents represent a lower share of GDP for countries in the region—an average of just over 5 percent during 1990 to 1994—than for comparable middle-income countries. Similarly, genuine savings are much larger relative to GDP than for other middle-income countries, although in this respect the Southeast Asian countries are slightly behind other East Asian countries.

East Asian countries also have been trying to finance very high levels of investment. This leaves a savings gap

that can only be financed either by foreign borrowing or by the depletion of natural capital (see figure 6.3). It is reasonable to assume that the crisis will restrict access to foreign capital flows, at least temporarily, and that there are limited possibilities to increase domestic saving rates, which have always been relatively high. Maintaining macroeconomic balance will, then, involve some combination of two options:

- Cutting the level of investment, which means either accepting a lower rate of growth or ensuring that investment resources are used more productively in the future; and
- Increasing the rate of natural resource depletion.

These are the key macroeconomic issues behind concerns about what may happen to natural resources during any economic recovery.

BOX 6.3

Economic growth or natural resource depletion?

Many countries in Southeast Asia have relatively large endowments of natural resources. These resources comprise a substantial share of total exports either via the direct export of minerals, fuels, foods, and raw materials or indirectly through processing industries that export a large share of their output.

This high proportion of natural resource exports may be viewed in two ways. The figure presents estimates of "genuine savings" as a share of GNP. Genuine savings is the difference between gross savings and the depletion of natural capital, or rather a measure of net savings for the economy after allowing for nonrenewable resource use. The results show that genuine savings as a share of GNP is consistently higher for both the Southeast and East Asian countries than for other middle-income countries. For the Southeast Asian countries, the share of genuine savings in GNP rose after 1980–84 and amounted to nearly 15 percent of GNP by the mid-1990s.

A second indicator is the "savings gap" as a share of GNP: the difference between gross investment and genuine savings (see figure 6.3). The savings gap measures the extent to which investment has to be financed either by foreign borrowing or by the depletion of natural capital. These may be regarded as interchangeable actions because both are equivalent to the creation of claims on future income in order to finance current investment. The depletion of natural resources means that rents from such natural resources will be lower in future, so income as conventionally measured will be lower. Foreign borrowing is a commitment to make repayments out of future income, thus lowering the net income available for domestic consumption.

The savings gap shows that the two groups of Asian countries have lower savings gaps than the averages for the groups of all lower and higher middle-income countries. The difference

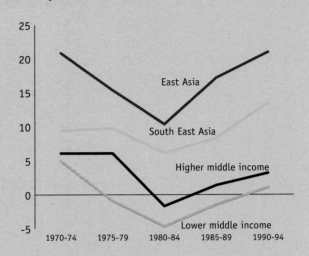

"Genuine savings" relative to GNP

Percent of GDP

between the Southeast Asian countries and the higher middle-income countries is slight, but it is much larger between the Southeast Asian countries and the group of lower middle-income countries. This reflects the extent to which all middle-income countries have come to rely more heavily on foreign borrowing or the depletion of natural capital to finance high levels of investment.

Source: McMorran and Hamilton, 1996.

Responding to the crisis

Pollution and economic growth. As a result of the financial crisis, many East Asian countries have experienced a fall in GDP in 1998 and will only recover gradually over the next two to three years. Lower levels of economic activity have reduced pollution from industry and vehicles, improving environmental conditions in industrial and urban areas. On the other hand, new investment, which is typically associated with less-polluting technology has slowed or ceased, causing the life of existing polluting industrial plants to be extended.

To examine these effects, projections of total emissions of key pollutants under "pre-crisis" and "post-crisis" scenarios have been prepared using a model that captures the effect of economic growth and

industrial change on output and emissions.[2] Discharges from small sources—small and medium industrial plants, vehicles, and households—have the most direct impact on average levels of exposure to pollutants. Hence, the analysis that follows will focus on trends in emissions from small sources.

The crisis is expected to have a significant impact on emissions of the main air pollutants from small sources. Figure 6.4 shows post-crisis projections for the two pollutants that cause the greatest damage to human health: particulates and lead. In addition, this figure shows the estimated changes in emissions due to the crisis. Many small sources emit particulates, largely as a result of burning various types of fuel, which provide the best indicator of air pollution in general. Lead comes primarily from cars using leaded gasoline, so lead particu-

FIGURE 6.3

The "savings gap" relative to GNP

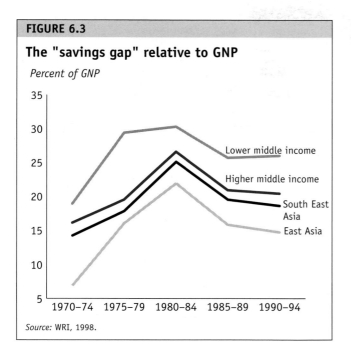

Source: WRI, 1998.

FIGURE 6.4

Small source emissions of key pollutants, Indonesia 1995–2020

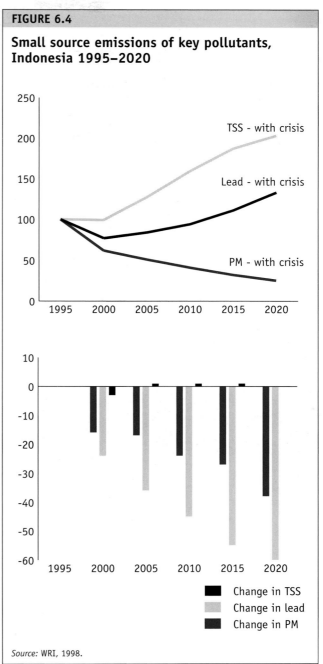

Source: WRI, 1998.

lates can be used as a general indicator of traffic pollution. The immediate impact of the crisis should be a reduction in emissions of particulates by the year 2000 by about 17 percent relative to the projected level had the crisis not occurred, and reduce lead particulates by about 20 percent. These reductions are based on the assumption that the growth that was lost during the crisis will not be recovered.

The story is somewhat different for total suspended solids (TSS), which are a general indicator of water pollution. As a result of efforts to reduce discharges of water pollutants by industrial plants and investments to improve access to water and sanitation, the level of emissions was projected to fall significantly by the year 2000 and in subsequent years. The crisis has had negative as well as positive effects because it will delay improvements in environmental performance. There will be a reduction in emissions of about 5 percent by 2000, relative to pre-crisis projections, but there will be a small increase in emissions from 2005 to 2015.

Economic recovery after 2000 will mean that emissions of air pollutants will soon exceed 1995 levels unless measures are taken to bring about large improvements in the average level of emissions per unit of GDP. In this respect, the medium-term impact of the crisis will be detrimental because investments will be delayed, which will, in turn, increase the average emissions per

unit of GDP by 5 percent to 10 percent in 2005. A similar concern about the impact of the crisis on environmental investments relates to the issue of access to water and sanitation, a critical environmental priority for most countries in this region. The average life expectancy of people in the Asia and Pacific region is shortened by nearly two years as a consequence of the lack of clean water and sanitation services.[3] Hence, it is

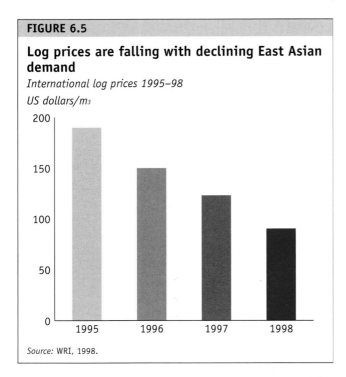

FIGURE 6.5

Log prices are falling with declining East Asian demand

International log prices 1995–98

US dollars/m₃

Source: WRI, 1998.

important to protect public expenditures allocated for this purpose, while in the long term, alternative sources of finance linked to improvements in service efficiency and greater cost recovery will be required.

Shifts in natural resource prices. Despite environmental concerns about the impact of currency depreciations on logging, the demand effects of the crisis have, so far, swamped any response to relative prices. South Korea and Japan, two of the largest importers of forest products, have reduced their demand for plywood by 30 percent. Indonesia's Minister of Forestry has predicted that the country's wood-related exports will drop by 25 percent in 1998, from US$8.3 billion to US$6.2 billion.[4] Many logging companies are in serious financial trouble. The *Jakarta Post*[5] reported that "at least 5.9 million cubic meters of cut logs remain in the forests because the timber estates have stopped operations."

Falling demand has reinforced the long-term decline in the world prices of logs and plywood (see figure 6.5), which will only reverse after a substantial shift in the balance between output and demand. New markets for timber products may open in response to policy changes or low prices. Even so, logging and production of wood products in the region are likely to continue to contract as a result of tighter constraints on credit and the availability of investment resources.

The long-term effects of the crisis will depend on the nature and extent of changes in relative prices. Forest management and land-use decisions rest on the relative values of both the capital stock and the flow of income over time. An increase in the absolute value of the timber stock need not alter the relative return from cutting timber now rather than at some time in the future. This would only be true if the real price of timber were expected to fall. Similarly, the returns from converting land from forest to plantation or agricultural land will depend on shifts in the relative prices of timber, perennial crops, and agricultural products. All are traded items; therefore, exchange rate changes should not favor one form of land use over others.

Nonetheless, the level and structure of user fees for exploiting natural resources should be adjusted. These have been consistently set at a level well below what would be justified by the resource rents. Increases in user fees would enable the East Asian governments to capture a more reasonable share of the income that is generated by logging, fishing, mining, and other natural resource activities. More important, the adjustments would provide an opportunity to establish a more appropriate structure of incentives for the proper use of natural resources.

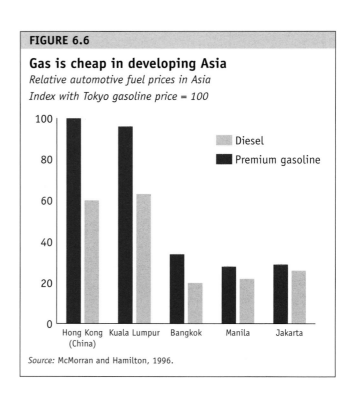

FIGURE 6.6

Gas is cheap in developing Asia

Relative automotive fuel prices in Asia

Index with Tokyo gasoline price = 100

Diesel

Premium gasoline

Source: McMorran and Hamilton, 1996.

A sharp increase in user fees would discourage any short-term tendencies to harvest stocks of natural resources in response to temporary swings in price relative to domestic goods as a result of exchange rate changes. However, compliance with such user fees, even at their current low levels, has been very poor; increasing fees without devoting more effort to enforcement will only achieve more distrust and a further loss of confidence in the overall system of natural resource management. It may be appropriate to phase in higher fees over a period of up to two years, provided that the required effort is devoted to improving collection rates. At the same time, regulations on harvest planning and oversight of concessionaires and other operators must be enforced more uniformly.

Eliminating subsidies. There are other distorted price incentives that result in significant environmental costs and should be eliminated as quickly as possible. In Indonesia, it is most important to eliminate subsidies and highly discriminatory taxes that favor the use of kerosene and diesel fuels over other fuels. In other Southeast Asian countries, transport fuels are much cheaper than elsewhere (see figure 6.6). There is substantial evidence showing that the fiscal impact of these pricing policies is regressive from a distributional point of view because they encourage the use of fuels that contribute heavily to local air pollution.

It may be argued that attempts to eliminate such subsidies have, in the past, provoked political unrest. However, public resistance should be seen in the broader context of fiscal policy and income distribution. Large increases in fuel prices or taxes may be difficult to implement on their own but are likely to be more acceptable if packaged with other tax changes whose overall impact is seen as being equitable. Among the issues that should be considered are:

- If taxes have to be raised or public expenditures reduced, is it better to increase fuel taxes or lower subsidies rather than, for instance, cutting back on expenditures that directly benefit the poor and those most severely affected by the crisis? It is important to distinguish between (a) resistance to changes in fuel taxes as a general protest against incompetent, corrupt, or inequitable economic policies and the (b) self-interest of those who benefit directly from distorted prices. For Indonesia, government subsidies in the 1998–99 state budget amounted to Rp7.4 tril-

lion (US$1.5 billion). It is difficult to reject the argument that such resources could be used more effectively in other ways to alleviate poverty and to mitigate the impact of the crisis.

- Trying to change fuel taxes and subsidies very quickly is usually doomed to failure. Progressive adjustments every quarter or half year will be more palatable, provided that the changes are signaled well in advance, so people have time to adjust.

- From an environmental perspective it is not critical that prices should be adjusted quickly. Short-run changes in the level and composition of fuel use in response to price are modest; most estimates of short-run price elasticity of fuel demand fall between -0.10 and -0.20. Over a period of one year, income changes are a much more powerful influence on consumption than prices. However, the medium- and long-run responses to price changes are much larger, typically four to five times the short-run elasticity as people change their habits, buy more fuel-efficient vehicles or equipment, and make the investments required to switch to different fuels. Large increases in relative prices may dramatize the need for change but may be counterproductive if they are subsequently reversed. Thus, a more gradual approach to which the government is fully committed is likely to be more effective.

Changes in public expenditures. To date, there is no evidence that environmental spending has been subject to disproportionate cuts. In the Philippines, for example, the budget of the Department of Environment and Natural Resources (DENR) has been affected by general measures including a 25 percent mandatory reserve on all expenditures other than personnel and debt service, a 10 percent deferral in the revenue allocation for local government units, and the suspension of tax subsidies to governments units. Various new programs and projects have been postponed, but the overall budget allocation, before mandatory reserve requirements, was increased in 1998.

Tight fiscal discipline will complicate a critical transition. It is essential to decentralize much of the responsibility for environmental regulation, especially in large, diverse countries such as Indonesia and China. Budgetary restraints may encourage central governments to delegate more responsibilities to regional and local levels, which will achieve little unless additional resources are provided, at least during the transitional period, to build up and sustain local administrative capacity. Thus, the pattern of government spending on environmental protection must change to reflect both new priorities and new approaches to environmental management.

Finding the resources to ensure that environmental regulation and enforcement can be effectively decentralized must be the most important goal to ensure that countries emerge from the economic crisis with a stronger capacity for environmental management.

Environment and poverty programs. Higher unemployment and lower real wages in urban areas have either reduced migration from the countryside to cities or reversed the flow temporarily. If this persists, pressure on rural resources, especially agricultural land and water, will encourage the conversion of marginal or forest land to agricultural land and may encourage rural migration to frontier areas.[6] Programs that provide employment opportunities, especially in rural areas, will mitigate the risk that such adjustments will undermine efforts to better manage natural resources.

Such programs will yield positive environmental benefits if money is allocated to improve rural infrastructure or to protect the environment. For example, public expenditures on rural water supplies and sanitation, planting trees, and soil conservation can generate substantial employment while improving the quality of life and/or productivity of rural populations. Similarly, expenditures on improving water supply, sanitation, and waste management in urban areas will help alleviate the poverty created by the crisis and will produce lasting environmental benefits for many urban residents.

A new path for the future

In periods of economic crisis, it is easy to assume that attention to environmental problems is a distraction from efforts to re-establish economic stability and growth. This presumption reinforces a widely held view in the Asia and Pacific region that environmental concerns are something that should wait until income levels are much closer to those in developed economies. The example of almost all industrial countries, including countries that many seek to emulate, such as Japan, seems to support this position.

Yet, developed economies did address some of the environmental problems associated with urbanization and industrial growth. They invested heavily in developing infrastructure for water supply and sanitation. Within the limits of the technologies available, they also tried to mitigate industrial pollution. In some respects, Asian countries are still well behind the achievements reached 80 to 100 years ago by well-developed economies.

Advances in knowledge and technology mean that the trade-off between growth and environmental quality also is very different. Often, the cost of reducing pollution is low or negligible, because modern production techniques and capital are much cleaner as well as more efficient. All that is required is to ensure that plants and equipment they are properly operated and maintained, something that is necessary if East Asian firms are to compete in world markets.

A simple continuation of past policies for the next 25 years would:

- Leave many households without access to clean water and decent sanitation;
- Worsen urban air quality in small and medium Chinese cities as well as in cities such as Bangkok, Jakarta, and Manila; and
- Increase the risks posed by heavy metals and persistent organic chemicals in rivers and water supplies but improve other indicators of water quality, such as suspended solids and dissolved oxygen.

Measures requiring an investment of less than 1 percent of GDP and with an annualized cost of 1 to 1.5 percent of GDP by 2020 will be sufficient to reverse the trends and improve the favorable ones.[7] The main expenditures relate to the goal of achieving full access to water within 10 years and to urban sanitation in 20 to 25 years, though in China the main goal is to deal with urban air pollution caused by burning coal for household heating and in small boilers. The benefits of these measures are five to 10 times higher than the costs involved for investments in water infrastructure and two to three times higher for expenditures on reducing air pollution.

Following this new path will involve different combinations of incentives, regulations, and institutional changes. However, the experience of economic crises elsewhere suggests that it may be difficult to achieve an appropriate balance when governments and the population at large are preoccupied by the immediate problems of either mitigating or adjusting to large economic changes. Thus, the key question is how to identify the steps that should be taken now to begin to establish a better framework for environmental management in the future.

For this purpose, it is beneficial to learn from the traumatic economic changes that have accompanied the transitions in Eastern Europe and the former Soviet Union. This transition will eventually lead to better environmental policies and conditions. However, the immediate economic costs have been so large in many countries that it has proved difficult or impossible to implement effective measures to correct the weaknesses of the previous regimes. The most rapid progress on the environmental front has been made by countries, such as Poland, the Czech Republic, and Hungary, that experienced the least economic disruption and were quickest to re-establish economic stability. There are four main lessons that emerge from these experiences:

- Re-establishing economic stability is an absolute priority even for economies concerned with improving environmental performance. Without economic stability, it will not be possible to obtain the support from the public and businesses that is required to implement effective measures to deal with environmental priorities.
- A clear public commitment to meet environmental and other goals that are consistent with practices in other countries or associations provides a reference framework for all of the agents involved in environmental management. This does not mean that countries should transpose European Union (EU) or U.S. standards. Rather, it would be better for the Association of Southeast Asian Nations (ASEAN) or member countries to commit themselves to the goal of developing environmental institutions and policies over the next decade that are mutually consistent and are broadly equivalent to those of OECD countries (allowing for differences in circumstances and resources). What matters is not the adoption of similar standards, but rather the development of reasonable mechanisms for agreeing and enforcing policies.
- Openness to trade with and investment flows from countries that have devoted more attention to environmental concerns is a powerful engine for the

transfer of better environmental practices without jeopardizing prospects for economic growth and increased standards of living. The frequent suggestion that liberalization of trade and investment will generate pressures to lower environmental standards is, in almost all cases, incorrect. There are many lessons that may be learned from investment and trading partners on how to improve both economic efficiency and environmental performance.
- Make haste slowly. Improving environmental performance will involve a commitment to transparency and decentralization in the formulation and implementation of environmental policies, such as in the banking, corporate, and public sector governance. This will require a fundamental shift in governance and will succeed only if addressed over a realistic time frame. It is easy to assume that the main objective should be to transpose the formal superstructure—technical standards, legislation, and regulatory mechanisms—of environmental policy. However, real improvements in environmental quality are the result of a broad consensus about environmental priorities and the measures necessary to improve the situation. There is no such consensus in most Asian countries. Thus, new governments must build public support for tackling a limited set of priorities before attempting to introduce and implement appropriate measures.

The financial crisis is only a transient event. The critical question for the environment is whether resumed growth will be "business as usual" or will reflect fundamental reforms in both the economic and environmental spheres. The linkages between the economy and the environment cannot be managed as directly as financial problems. Solutions require the strengthening of regulatory, institutional, technical, and managerial capacity with an emphasis on cross-sectoral coordination and consensus. Building such capacity and fostering change will require a prolonged effort and proper incentives.

Environmental improvement in the region cannot be financed by the government. It is the job of the government to establish rules and a clear regulatory framework as well encourage those enterprises and agents with the maximum scope to achieve environmental goals. In turn, this means that the nature of environmental policies must change. There should be less focus on emission standards that promote the adoption of end-of-pipe pollution controls and more emphasis on pollution prevention combined with the adoption of cleaner and more efficient techniques of production.

In the immediate future, companies must be encouraged to achieve the best results by modifying their existing facilities. Two sets of incentives will tend to reinforce this action. The large exchange rate adjustments will mean that the cost of new controls will be more expensive than before the crisis, so firms should prefer to reduce emissions by adjusting their operational practices, training staff, and ensuring that their plants and equipment are properly maintained. Furthermore, pollution represents a waste of raw materials and other inputs, so there will be strong economic reasons to minimize such waste.

In countries where state controls over the energy sector mean that pricing issues are hotly disputed, there is great resistance to adjusting fuel prices in line with exchange rate changes. Inflation is not especially sensitive to changes in the overall price of energy, whereas the subsidies to hold down all or specific fuel prices are almost invariably regressive. In fact, those subsidies are often simply a matter of special interests attempting to protect valuable privileges.

Among the most important of those interests are energy-intensive industries whose profligate use of energy is almost always linked to heavy pollution. Although there may be political constraints on rapid adjustments in the prices paid by households for certain widely used fuels, there is absolutely no reason to protect industrial and commercial users. Thus, at the very least, the wholesale prices of diesel fuel, heavy fuel oil, gas, and coal should be brought into line with world prices within one year.

Public spending on basic water and sanitation infrastructures has an immediate return in terms of the welfare of low-income populations and will lessen the impact of the economic shock on employment. The costs of the lack of access to clean water in rural areas represent a large fraction of the total health damage associated with environmental factors. Hence, there is a strong case for focusing spending on rural employment, social assistance programs, and investments in the water supply infrastructure. The willingness to pay for clean water is high, so it should not be difficult to establish mechanisms to ensure that the operational costs of new systems are fully covered by modest user charges.

Notes

1. World Bank, 1992.
2. The model is based on an input–output framework with separate matrices of coefficients for old and new capital (the coefficients for old capital change gradually over time). The basic assumption of the model is that over the next two decades, less-developed economies in Asia will converge toward economic performance and industrial structures similar to middle-income countries. It is assumed that East Asian countries will gradually adopt efficient technologies and, as a result, energy and material inputs per unit of output will decline.
3. World Bank, 1997.
4. *Jakarta Post,* December 30, 1997.
5. *Jakarta Post,* January 15, 1998.
6. Cruz and Repetto (1992) argue that such migration was a consequence of unemployment resulting from IMF stabilization programs in the Philippines during the early 1980s. This interpretation is controversial because there were many other distorted incentives encouraging agricultural expansion in upland areas and the trend was well established before unemployment rose.
7. World Bank, 1997.

chapter seven

Priorities for a Sustainable Recovery

The ruins of crony banks and businesses built on corruption and special favors litter the devastated economies of Southeast Asia, making it easy to notice what went wrong. But buried under the rubble are a number of things that these societies did right—steps they took, or societal strengths they preserved, that may soften the trauma of the economic crisis they are going through. These values and accomplishments may well serve as a cushion on which the societies of Southeast Asia can bounce back once the economic panic recedes. —New York Times, May 31, 1998.

East Asia is in its second year of economic crisis. The fires of instability are almost contained in some countries, but are far from being under control in the region as a whole. Not only could they erupt anew in any country, they still threaten to sweep into other emerging markets. Although it is still too early to predict their final course or when they will be extinguished, three facts emerge with stark clarity:

- The level of devastation—in wealth losses, lost economic output, and in peoples' lives—is severe. Tens of millions of people are likely to be pushed below the poverty line and for millions more the climb out of poverty will be arrested, at least temporarily.
- The crisis has taken on systemic proportions in Thailand, the Republic of Korea, Indonesia, and Malaysia. In these countries, many banks and firms have been forced into insolvency, and many more are hovering on the brink. The steady compounding of unpaid interest onto the balance sheets of the banks and corporations has created a growing mountain of debt. The remedy will go far beyond a macroeconomic adjustment to accommodate the

reversal of capital flows. It must extend to the microeconomic and institutional restructuring of entire economies.

- The recession has become regional in scope, making it difficult for any one country, no matter how effective its policies, to escape the pressure of downward forces solely on its own. One country's potential export response is dragged down by its neighbors' import contractions. Credit contractions in domestic financial markets are magnified by credit contractions in the financial markets of neighbors.

These events, left unchecked, threaten to engulf the whole of East Asia and could even imperil the otherwise robust expansion of the world economy. Whether this comes to pass will depend on the way the region and the international community responds.

The stakes are high. Output in the East Asia 5 has dropped sharply in the first half of 1998. Because current account balances have swung from negative US$27 billion in 1997 to a projected US$40–50 billion surplus for 1998—a whopping 7–8 percent of gross domestic product (GDP)—at a time when overall output was falling, the compression of consumption and investment has been even more brutal. For the remainder of 1998, the consensus projections[1] paint a bleak picture: GDP is projected to decline by 16 percent for Indonesia, about 4.7 and 7.9 percent for Korea and Thailand, respectively, and by 3.4 percent for Malaysia. Even Hong Kong (China) may contract by as much as 3 percent. Of the East Asia 5, only the Philippines will record marginally positive growth. The smaller economies of the region—from Mongolia to Papua New Guinea and the Pacific Islands—are experiencing economic contractions. Consensus forecasts are still strong for China (7.2 percent), Vietnam (6.1 percent), and Taiwan, China (5.2 percent), but each is being pushed below its trend-line path.

This grim reality in the contracting economies has been accompanied by falling wages, rising unemployment, a shift of labor from high-wage to low-wage jobs, and sharp cuts in average per capita private consumption. In Thailand, for example, rice export volume is up 76 percent in the first quarter of 1998 relative to 1997, probably indicating that people are working harder and eating less. The economic expansion that raised the incomes of the poor and effectively provided a social safety net has ended, leaving large segments of society

vulnerable. Not surprisingly, the poor and politically disenfranchised, especially poor women and children, ethnic minorities, and migrants, are suffering the most. Though formal studies are just beginning, media accounts and interviews corroborate this conclusion. The El Niño-induced drought and forest fires, in addition to and increased poverty have put added pressures on the environment, although, ironically, the recession is forcing conservation that may countervail these negative effects.

The most urgent task ahead is to restore the conditions for robust economic growth throughout the region, particularly in the most adversely affected countries of Thailand, Korea, Indonesia, and Malaysia. If output were to fall by a cumulative 10 percent over the next three years and income distribution worsens by 10 percent, the number of poor people in Indonesia, Thailand, Malaysia, and the Philippines would more than double—from approximately 40 million to over 90 million. For these reasons, establishing the conditions for renewed growth is essential to reverse recent trends toward falling wages and growing unemployment. Of the other developing countries in East Asia, China alone has some latitude to chart its own course, because of its size, credibility of its financial policies among its citizens, and semi-closed capital account. But, it too will have to work hard to avoid an excessive slowdown in its high growth momentum, as will all of the smaller countries.

Restoring sustainable growth throughout the region will not be easy. Recovery hinges upon: revitalizing the banking and corporate sectors; reactivating aggregate demand; maintaining the pace of the structural reform agenda; ensuring low-income groups are protected during crisis and participate in the eventual recovery; and restoring international capital flows. These issues are inextricably related and require coordinated, consistent approaches. Because countries confront quite different domestic economic circumstances, policy priorities within the agenda will vary among countries.

Dealing with systemic banking and corporate insolvency

Throughout the region, the financial and corporate sectors are trapped in debt. Except for China, highly leveraged corporations are caught between falling revenues and rising costs for interest service and imported materials. Recession forces corporations to delay or default on bank payments, and, as the amount of non-performing loans (NPLs) rises, banks' cash flows are squeezed. This forces the banks to contract new lending to illiquid corporations and call in good loans to raise cash, further deepening the recession. In addition, tight monetary and credit policies have induced precipitous declines in the volume of real credit, impeding corporate revival. To be sure, authorities' choices were limited, and tight credit policies may have prevented the volume of financial savings from falling yet further. In any case, credit is scarce and many borrowers hover on the brink of default, so banks are unwilling to lend to all but the most reliable borrowers. Getting accurate information on firms and financial entities is difficult under conditions of exchange rate instability and frequent variations in relative prices; this complicates the process of differentiating the good borrowers from the bad, and underscores the importance of exchange rate and price predictability.

In Thailand, Korea, and Indonesia, the abrupt change in economic conditions has produced a systemic crisis. Rising debts, plus capital losses associated with exchange rate depreciations, have pushed a large segment of the countries' banks and corporations into insolvency simultaneously. According to simulations in chapter 4, in Thailand, one out of four listed firms are estimated to have balance sheet losses greater than equity. When this is combined with falling demand, NPLs are estimated to range from 20 to 35 percent of total loan portfolios. In Indonesia, two out of three listed large firms are bankrupt according to this criteria, and non-performing loans may reach as high as 50 percent. In Korea, two out of five corporations have exchange rate and interest rate losses greater than equity, and NPLs also range from 20 to 35 percent. Balance sheets and cash flow positions for the corporations are deteriorating to such an extent that, unless restructuring and debt workouts are carried out up-front, even a relaxation of monetary and fiscal policies is unlikely to produce the desired impact on corporations' finance and operations. Each month that passes worsens the situation because interest costs continue to mount.

Countries farther from the epicenter of crisis are faring better. In Malaysia and the Philippines, only 5 and 16 percent of firms, respectively, have losses associated with the shock greater than equity, and their payment record on debt is correspondingly better. In these countries, as with some of the small countries, restructuring may be able to proceed with less government coordination of the workout efforts. In China, the banking sector is also plagued by a high ratio of nonperforming loans (ranging from the official estimate of 20 percent to private estimates of up to 40 percent). State enterprise profits have plummeted not because of external factors, as in the East Asia 5, but rather in response to years of progressive exposure to new competition from the nonstate sector. Tight credit policies after 1994 have induced reforms, but many state companies are falling behind in their debt service to banks. At present, the problem is large but manageable for several reasons: (a) depositors believe in implicit government guarantees, and thus, the system is not exposed to panic-driven destabilizing outflows, and (b) the government has both the assets and borrowing capacity to finance a resolution. The government has already adopted a limited recapitalization of state-owned commercial banks and implemented a new system of prudential regulation. But it has to maintain a brisk pace of reform of both state enterprises and banks—even in the face of considerable external economic uncertainty—to avoid systemic problems as the capital account gradually becomes more open.

The pace at which countries reactivate credit flows to real economic activities will influence the speed of economic recovery. In the financial sector it is urgent to renew credit flows to exporters and to new investment. To reverse the microeconomic origins of high interest rates—eroded corporate balance sheets, deflated collateral values, and banks' aversion to accept risk when they desperately need to rebuild capital—will require time and a healthier economy. This will probably take years rather than quarters (see box 7.1).

Prompt recapitalization of banks up to international standards, and enhanced monitoring of these would help reduce the credit crunch and reduce uncertainty. Applying forbearance to stated capital adequacy targets is not a wise policy. Most countries only saw their problems increase when they allowed financial institutions to violate prudential rules. Capital adequacy targets should be meaningful, however, and be made in line with the projected profitability of banks and the

BOX 7.1

How long must East Asia wait to recover?

A recent study attempted to estimate costs of lost output for banking and currency crises by comparing GDP growth after a crisis with trend GDP growth for a group of more than 50 countries during 1975–97 (WEO, 1998). The study indicated that the cumulative loss in output per episode for banking and currency crises averaged some 14–15 percent of GDP. Average recovery was shorter in emerging market economies (2.6 years) than in industrial countries (5.8 years) though the cumulative output loss was larger, on average, for emerging market economies (18.8 percentage points) than for industrial countries (17.6 percentage points).

Argentina (1981), Chile (1982), Sweden (1991), and Finland (1991) offer useful comparisons. Five years before and after the crises, output growth, investment/GDP, and consumption/GDP ratios in these countries displayed common patterns of recession and recovery (see figures below). In the years running up to the crisis, the upward trend of investment/GDP ratio provides some support for the widely discussed boom-and-bust explanation of financial crises. As the crisis ensued, the bulk of the adjustment fell on the investment ratio, which declined dramatically. If gross national savings decrease owing to expansionary fiscal policy and bank restructuring costs, and if countries can no longer run current account deficits because of lowered capital inflows and reserves, the investment ratio must decline. This adjustment

occurred in most of the previous financial crises and has begun in Asia's crisis-stricken countries. Indeed, as a consequence of the increase in incremental capital output ratios (see chapter 1) and significant investment in low productivity areas, adjustment in investment in the face of a much higher cost of capital will be inevitable.

Investment to GDP ratio and financial crises

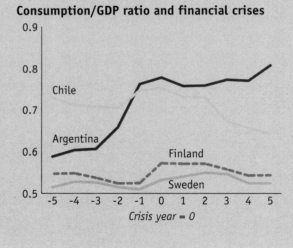

Real GDP growth and financial crises

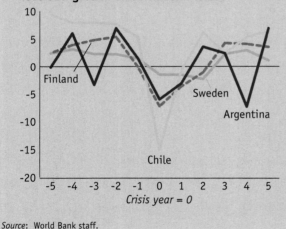

Crisis year = 0

Consumption/GDP ratio and financial crises

Crisis year = 0

Source: World Bank staff.

amount of public support that is available. Banks may also be given some tax relief, importantly on the provisioning for non-performing loans, so as to increase their retained earnings, and thus boost their capital. But, when announced, banks should be held to standards, including through rigorous portfolio audits, and forbearance minimized. In the absence of sufficient public support, it may be more effective to target support to a

few, institutionally well-developed banks which are kept to high standards, than to spread support over a large number of financial institutions.

Ultimately the banking system is only as healthy as its corporate borrowers. In the short term, two measures can help the corporations faced with financial restructuring. Providing some *(temporary) tax relief to corporations* would improve their cash flow and credit standing. Also, providing limited guarantees for *credit for collateralized transactions* can alleviate aspects of the liquidity crunch.

In several countries, policies to encourage *voluntary restructuring of corporate debt* are now gaining primacy as an alternative to court-supervised bankruptcy proceedings. These include eliminating tax disincentives to equity restructuring through mergers and acquisitions. In Indonesia and Thailand, for example, an asset transfer is treated as a taxable transaction; however exemptions should be allowed if the corporate seller does not receive cash. Similarly, removing tax disincentives to debt restructuring would speed voluntary restructurings. The government should also encourage conversion of debt to equity as part of voluntary reorganizations. Moreover, liberalizing treatment of foreign direct investment in selected sectors where it is now restricted and in the ownership of real property could open up new sources of management and capital. In Korea, some important steps have already been taken by the government to liberalize foreign investment (in both corporations and financial institutions), clarify Mergers and Acquisitions' rules, provide tax incentives and clarify tax and prudential treatment of restructured debt. More recently, the government has taken several steps toward allowing takeovers by foreign investors, moving toward 100 percent foreign ownership, allowing foreign investment in previously restricted business areas, and liberalizing the real estate market, including to foreigners. Finally, the government probably will have to help *establish a framework* for collective yet voluntary negotiations between debtors and creditors, including providing the criteria and financing to permit companies access to foreign exchange at predetermined rates to service foreign debts. Most East Asian countries have yet to complete the formulation of a comprehensive framework of corporate sector restructuring. Nevertheless, crisis countries like Thailand, Korea, Indonesia, and Malaysia have adopted a "London Rule" type, voluntary approach to help viable companies restructure their debts. The government can also allow regulatory authorities to treat new lending to companies that have undergone restructuring as more secure than *existing debts for purposes of provisioning* thereby creating an incentive for firms to restructure and for banks to lend. Virtually all countries have

revised their *bankruptcy statutes* that will, among other things, provide a credible impetus to voluntary restructurings. For example, Thailand, Korea, and Indonesia have introduced amendments to their bankruptcy laws to strengthen the capacity of the courts to approve reorganization proceedings and to supply specific provisions and rules that reduce the discretionary power of the court and increase the degree of transparency, certainty, and efficiency in court proceedings.

While these measures may stave off bankruptcies in a fundamentally viable but illiquid system, they cannot salvage a corporate and financial sector rendered technically insolvent by the new exchange rate/debt burden combination. If, under the new constellation of exchange rates, the present value of corporate assets is still less than the present value of liabilities, pervasive insolvencies will threaten to bring down the financial system. In these circumstances, the government has no alternative but to play a strong leadership role to quickly restructure the financial and corporate sectors. Like Chile in 1982 and Mexico in 1994, Asian governments may have to intervene to ensure the survival and restructuring of both the financial and corporate sectors. The Chilean and Mexican governments made depositors and lenders whole, restructured the companies under new ownership, and auctioned off bad portfolios and the newly restructured companies to preserve growing concerns. This decision requires governments to balance the burden on governmental management capabilities against market-based reorganization and to select the option that will produce the fastest recovery.

The history of restructuring offers some hope that this process need not be a prolonged drag on recovery. Chile undertook massive restructuring and, after only three years, began its longest sustained expansion in this century. Successful restructuring experiences around the world point to certain imperatives:[2]

- *Conduct an early and systematic evaluation* to design a strategy (including differentiating viable banks from those that are clearly bankrupt), and then act immediately.
- *Adopt a comprehensive approach to repair* the immediate stock and flow problems of weak and insolvent banks; repair the shortcomings in accounting, legal, and regulatory frameworks; and redress weak supervision and compliance.

- *Transfer nonperforming loans* from the banks' balance sheets to a separate loan recovery agency to alleviate the banks' stock problem.
- *Provide capital* to viable banks during restructuring.
- *Set up mechanisms to compel operational restructuring* so as to return banks and firms to profitability and sustained solvency.
- *Enforce exit policies for firms* to allow owners to declare bankruptcy, and sell assets so they can be put back into production, albeit with reduced book values.
- *Require loan workouts* to recover some of the costs of bank restructuring and to send signals to delinquent borrowers.

A corollary problem is the restructuring requirements posed by public sector companies in the region. Power companies, water utilities, and other state companies have suffered the same interest and exchange rate shocks as private firms. On the one hand, this increases demands on public finances already strained by new burdens. But on the other, the crisis has created the opportunity to rethink the disposition of these public assets, the efficiency of public monopolies, and ways to leverage new private capital into these sectors. Governments should not hesitate to review the benefits of privatizing state companies and reorganizing these sectors to permit new entry from private firms.

Privatization is not a panacea, and indeed the efforts to mobilize private participation in some sectors have created new problems, as contingent liabilities in contracts with private producers have created increased new public debts because of exchange rate changes. Nonetheless, in many cases, governments can achieve the twin objectives of garnering new capital revenues for the state and provide services more efficiently through partial or complete privatization.

The need to restore growth in aggregate demand

Demand is depressed. At home, plummeting stock and real estate prices brought about dramatic net-worth losses for households and businesses. Middle class savings have been decimated, and economic contraction has universally shrunk purchasing power. Also, the loss of investor confidence has sharply reduced investment demand. Meanwhile, foreign demand has so far not compensated for falling domestic demand.

It is useful to preface the discussion with an assessment of the relative magnitudes of the components of demand, and their recent role in the region's growth. Two factors distinguish East Asia—particularly the East Asia 5—from other developing countries: very high investment rates, plus a large share of exports in

TABLE 7.1
Contribution to GDP growth (1990–96)

	Private consumption	Investment	Government consumption	Net exports	Exports	Imports
Indonesia						
Contribution to growth	62.5	31.4	7.1	-1.0	27.8	-28.8
Share in GDP	57.7	33.0	8.6	0.8	27.0	26.2
Korea, Republic of						
Contribution to growth	55.3	40.1	11.3	-6.7	31.1	-37.8
Share in GDP	53.6	37.0	10.5	-1.4	30.2	31.6
Malaysia						
Contribution to growth	42.7	49.6	9.7	-2.0	102.5	-104.5
Share in GDP	49.9	37.9	13.0	-2.5	86.5	89.0
Philippines						
Contribution to growth	78.0	26.9	12.9	-17.8	47.4	-65.2
Share in GDP	73.9	22.8	10.6	-6.6	32.8	39.4
Thailand						
Contribution to growth	52.0	44.4	10.7	-7.0	41.3	-48.3
Share in GDP	54.7	41.1	9.8	-5.5	37.9	43.3

Source: IFS.

The Pacific Islands: Bending in the winds

Pacific Islanders, long familiar with natural disasters, such as cyclones, droughts, tidal waves, or volcanic eruptions, like to compare themselves with the coconut palm that bends with the wind but can quickly put out new growth when the storm has passed. The impact of the East Asian financial crisis has tested this resolve.

Growth was already weakened by severe El Niño-related drought, falling commodity prices, reduction in yields of key crops, the social impact of civil service reforms, political instability, and a drop in the level of overseas development assistance (ODA). Ironically, the three largest economies—the ones with the greatest potential for economic security, Papua New Guinea, Fiji, and the Solomon Islands—have been hardest hit to date.

- The loss of export commodity markets for logs, copra, and minerals in East Asia led to significant exchange rate depreciations in late 1997: the Kina (PNG) dropped by about 30 percent and, the Solomon Islands and Fiji devalued their currencies by 20 percent.
- For the Solomon Islands, goods exports may decline by about 20 percent while real GDP may decline by 10 percent or more in 1998; for PNG, the adverse impact is in the range of 2.6–4.8 percent of GDP. Fiji has revised its forecast to a negative growth rate for 1998.
- The reduction in demand from Asia for other commodities will have longer-term impacts on the smaller economies. For example, markets for Vanuatu beef and tourism, Nauru phosphate, Tongan squash, and Kiribati fish have all significantly weakened.
- Tourism has declined dramatically, particularly in Micronesia, (FSM, RMI, and Palau) Fiji, and Vanuatu, and several airlines servicing Asian/Pacific routes have withdrawn their services. Hotel investment is also facing a setback.

Apart from currency devaluation (in the Solomon Islands and Fiji), governments have tried to continue economic reforms already in place and to renew regional dialogue on strengthening trade relations. There is also a greater interest in securing more favorable market access in ongoing trade negotiations and in simplifying national investment policies.

Source: World Bank staff.

GDP. Since 1990, on average, investment spending accounts for almost 40 percent of GDP growth in the East Asia 5. This implies that in the first phase of adjustment, contraction could fall heavily on invest-ment, easing the pressures on consumption; the flip side of the coin is that it will have to play an important role in recovery. As discussed above, this will be difficult as long as the financial and corporate sectors are struggling to restructure their balance sheets and real credit growth is paltry. Similarly, exports have accounted for a large growth impulse—30 percent or more—offset in recent times by growing imports. This suggests that, if regional and world market conditions were favorable, and the region lowers its import intensity in response to currency depreciation, trade could play a more dynamic role in East Asia than in comparable crises elsewhere. It also suggests that, since investment demand is constrained by bank restructuring and consumption by falling incomes, we should look carefully at net exports and government spending.

External demand: The regional export trap

Although net exports have tended to be a drag on growth in the recent years of increasing current account deficits, they could play a leading role in recovery provided the current robust global market conditions remain. Indeed, the crisis in East Asia would have been more severe if the global environment had not been strong.

Export performance has so far been disappointing. The regional tailspin has led to export price declines that have eroded much of the value of export volume increases, though perhaps less so for the Philippines and China. In the first quarter, export volume increases exceeded 20 percent for much of the region, but the fall in unit prices by about 12 percent has cut their increase in value to the exporting economies to a mere 8–10 percent. To the extent export increases are based on inventories sold off to cut working capital costs, the effects on real economic activity may be overstated. The current account adjustment has been massive: a swing of US$80–90 billion in one year. However, it is the domestic economic contraction that has led to a sharp reduction in imports. Because some 40 percent of trade is intra-regional—50 percent if Japan is included—the effect of this contraction has been shared deflation. This region-wide recession makes it more difficult for any one country to recover. The pace of export recovery therefore depends primarily on the speed at which exporters can shift their potential markets from inside to outside of the region, and to secondary demands generated within the region.

The growth of markets outside of East Asia holds one of the keys to the region's recovery. Growth in Europe is strong, despite an apparent peaking in the U.S. business cycle—a fortuitous synchronization. Inflation remains low and commodity prices have been declining, allowing monetary authorities in the United States and Europe to pursue less restrictive policies. Except for Asia, world trade growth continues to be robust.

Unless the crisis spreads, the shock of the Asian crisis is unlikely to have a large adverse effect on the world economy. Even though East Asian demand for imports has fallen roughly 30 percent, this accounts for only 0.3–0.5 percent of U.S. GDP. The fundamentals supporting U.S. growth remain sound while European growth is strengthening; in Germany and France industrial production is rising; in Italy and the U.K. growth is somewhat less. Unemployment remains high but is on a downward trend in France and Germany.

World trade volume grew by 9–10 percent in 1997, then slowed to around 8 percent in the first quarter (year on year). Nonetheless, world trade is now projected to grow by about 5–6 percent in 1998 and at about the same pace in 1999, barring any further unraveling in East Asia or other major developing countries. Yet three questions hang over this outlook.

First, will Japan's domestic demand recover in the next two quarters? Japan contributes about 50 percent of total regional GDP and absorbs a 15–20 percent share of the exports from other East Asian countries. However, a significant part of this demand comprises intermediate inputs to products destined for the United States and the European Union (EU), meaning that Japan probably accounts for less than 10 percent of final demand, and that the United States and the EU could account for up to 80 percent. Nonetheless, Japan is a competitor to some East Asian countries, notably Korea and Taiwan (China) in some product lines. The yen depreciation multiplies competitive price slashing because it not only reduces the dollar-based costs of imported Japanese inputs, but also intensifies competition from Japanese exports. Unless there is a Japanese

recovery that bolsters the yen, the depressing effects on East Asian export prices are likely to remain severe.

The Japanese economy also influences world trade volumes and prices in selected commodities. Japanese demand, together with the rest of the region, affects oil prices, and thus exports of Indonesia, Malaysia, and Singapore; Japanese rice purchases affect Indonesia and Thailand; and Japanese rubber demand affects Malaysia and Thailand. To some extent, falling demand in the region has depressed commodity prices, so Japan plus the rest of the region hold the key for these prices.

Japanese performance also influences financial markets. Japanese banks have US$271 billion of credit external outstanding (mid-1997)—more than half in Hong Kong (China) and Singapore and most of the remainder in Korea, Thailand, Indonesia, and Malaysia. Japan's foreign direct investment (FDI) flow to the region was about US$11 billion in FY1996, mainly to China, Hong Kong (China), Singapore, and Indonesia. Japan is an important market for the rest of East Asia's exports as well—from a high of 27 percent for Indonesia to a low of 5 percent for Hong Kong (China).

Japan has already enacted a stimulus package of US$122 billion or (3 percent of GDP)—half in public works spending and half in tax cuts. In addition, it has announced plans for supplemental spending of US$70 billion and a permanent tax cut of US$45 billion. The expansionary impact will probably be somewhat counteracted by lack of consumer confidence that will increase propensities to save. Also, the placement of bonds to finance the stimulus may raise long-term interest rates and attract capital inflows, fortifying the weakened yen. Monetary policy could offset these effects to ensure that recovery would take hold. If a Japanese recovery gains traction in the latter part of 1998, these developments would markedly improve the outlook for East Asia.

A second question is whether the U.S. policy stance will be changed. Emerging trade deficits with the region may generate demands for trade restrictions or other policy response. The U.S. trade deficit has worsened significantly as cheap East Asian products have entered U.S. markets, producing a bonanza of cheap products in stores everywhere. These have also dampened any inflationary pressures. The decline in equity markets in

BOX 7.3

Vietnam: Learning from the misfortunes of others

Vietnam's economy has yet to be fully integrated into global financial markets and its capital account is tightly controlled. Therefore, initially, policymakers felt that the Vietnamese economy would be insulated from the negative ramifications of the crisis sweeping the region, which led to complacency about the need to reform. At first, complacency seemed to be a successful strategy: economic growth reached nearly 9 percent in 1997; inflation remained low at less than 5 percent; exports grew at a reasonably healthy pace compared with other economies; and foreign direct investment inflows were US$2.6 billion, higher than in 1996. As the crisis has deepened, however, growth has slowed significantly and could lead to other setbacks for Vietnam.

Declining international price competitiveness. Since July 1997, the currencies of Thailand, Indonesia, the Philippines, Malaysia, and Korea have depreciated substantially. Over the same period, Vietnam's real effective exchange rate has appreciated. This trend has had two effects: it has adversely affected the relative price competitiveness of Vietnam against other ASEAN economies; and has led to some speculative pressure on the Vietnamese Dong, despite its nonconvertibility.

Declining export demand and prices. Economic growth in Asia has slowed dramatically. Domestic demand has collapsed in the major economies of ASEAN. As a result, export demand will be lower. This trend is affecting Vietnam, as in 1996, about 40 percent of Vietnam's exports went to ASEAN and another 20 percent was absorbed by nonASEAN Asian countries. Prices of Vietnam's major export commodities have declined and this trend also is

having a significant impact on export revenues. While prices of some primary commodities have revived somewhat recently, they are likely to remain weak. Recent gains in rice prices, for instance, have been more than offset by declines in crude oil prices, driven partly by declining demand in Asia. Prices of imported Asian textiles and garments in Europe, Japan, and the United States have also declined in recent months. In the first six months of 1998, exports grew only 10.6 percent in U.S. dollar terms, compared with 30.6 percent in the previous year.

Declining foreign direct investment. Foreign investment has been an important source of productivity and employment growth in Vietnam in the 1990s. Asian economies were the largest investors in Vietnam and, as their economic fortunes have plummeted, their foreign investment there has fallen sharply. In 1998, both FDI inflows and approvals have dropped abruptly; in the first 6 months FDI inflows totaled only US$1.0 billion compared to US$1.7 billion in 1997. FDI approvals also have fallen dramatically, from US$1.6 billion in the first 6 months of 1997 to US$ 1.2 billion in 1998. The primary reason has been the sharp adjustment in economic activity in Asian economies as a result of the crisis. Vietnam's decline in competitiveness may also entice foreign investors to other countries. Finally, to the extent that foreign investors reappraise their views of the region in general, foreign investment inflows will fall.

Source: World Bank staff.

the late summer of 1998 raised concerns about a slowdown in the U.S. economy, and ignited discussion about a more expansionary monetary policy. The outcome of this discussion will have considerable import for East Asia, because the continuation of the strong U.S. expansion is essential for an East Asian recovery.

A third question is what will happen in other large emerging market economies that may be vulnerable to currency or financial crises? There is already turbulence around Russia. Rumbling in the background are concerns about an economic slowdown in China that might force a depreciation of the renminbi—an event widely thought to have potential to jeopardize the Hong Kong (China) currency board and to undermine the Taiwan (China) dollar.

The Chinese government has indicated that it would not depreciate its currency unless forced to do so by external events. China has launched a package of monetary and fiscal measures to stimulate the economy, and

internal demand far outstrips any possible marginal contribution from enlarging its already massive trade surplus. The government also is pushing forward with financial reforms to strengthen its domestic banking system. At present, the country enjoys a strong reserve position, low short-term debt, and has a semi-closed capital account.

Contrary to popular perceptions, there are no technical reasons why a depreciation in China would spell disaster for the rest of the region. Any depreciation is unlikely to be large, trade and financial links with the rest of East Asia are relatively small, and global models confirm that a depreciation would only minimally affect neighboring countries. However, models do not capture expectations or the unpredictable herd behavior of financial markets.

This discussion suggests three conclusions. First, export receipts are likely to remain anemic due to weak intra-regional demand and prices—unless growth can be jumpstarted simultaneously. This situation could persist for several quarters, or at least until the region as a whole either picks up some growth momentum or exporters establish new markets, or both. Second, once exports do pick up, the upside potential is great because exports play such a large role in the economies of the region. In the meantime, East Asian countries must avoid competitive depreciations, which benefit no country and bring monetary instability to all. In addition, excessive depreciations would further undermine corporate distress by raising funding costs. Imposing import restrictions to improve balance of payments should also be avoided as it would starve domestic production of critical inputs and lead to a decline in production. Countries may, in fact, wish to accelerate the pace of trade barrier reduction already agreed to under APEC arrangements.

Government demand: The dilemma of adjustment

Although government consumption demand in conservative East Asia has historically constituted less than 10 percent of growth (table 7.1), short-run fiscal policy is a potent instrument to influence aggregate demand. Indeed, considering the vast social needs ensuing from the crisis, direct government expenditure programs can alleviate the suffering of the poor and unemployed, while boosting demand in the corporate sector. Already, the cost of servicing liabilities taken over from the financial sector, together with falling revenues from the shrinking tax base, have plunged public sectors into deficit.

From the onset of the crisis through early 1998, fiscal policies, contrary to their design and with benefit of hindsight, turned out to be contractionary. The fiscal positions were not designed *ex ante* to be contractionary, but assumptions about economic growth in the early programs, though broadly in line with private forecasts at the time, proved to be too optimistic, so on a cyclically adjusted basis, *ex post* fiscal positions proved contractionary. For example, in Thailand, the August 1997 program contemplated a tightening of 3 percent of GDP to control aggregate demand and to pay for the costs of the financial sector workout, and was predicted upon a growth assumption of 2.5 percent growth in output. As it happened, the economy significantly worsened.[3]

As the depth of the recession has become clear, all governments have loosened their fiscal stance, with the active support of the IMF (with much of the spending destined to the social sectors). Table 7.3 shows the objectives of the region's governments for the central government balance (before privatization revenues and financial sector changes). Using 1995 as a full-employment base year, it also shows the fiscal stance relative to a neutral path of revenues and expenditures had the economies continued to perform at near full capacity. Finally, it shows the fiscal impulse relative to 1997. (A positive sign indicates a net stimulus and a negative sign indicates a contractionary impulse.) Even though in Korea, the Philippines, and Malaysia, as late as the first quarter 1998, fiscal targets were contractionary relative to 1997, fiscal policy has since become expansionary in all countries.[4] This has removed a brake from aggregate demand within the region.

It remains to be seen whether this fiscal stance will be enough to reactivate these economies. The situation is different in each country. Korea has more scope for additional stimulus than other countries because of its

TABLE 7.2
Moving targets: East Asia's changing fiscal stance
Fiscal objective for 1998 in second and third quarter programs

	Thailand		Korea, Rep. of		Philippines		Malaysia		Indonesia	
	February	August	February	August	March	July	April	July	April	July
Government balance	-1.7	-2.7	-0.7	-4.3	-1.0	-1.8	3.1	-3.5	-3.2	-8.4
Fiscal stance	2.9	3.4	-0.1	2.9	-0.5	0.2	-0.2	3.9	2.4	6.2
Fiscal impulse	0.0	0.5	-0.8	2.2	-0.1	0.6	-0.5	3.6	1.2	5.0

Note: Excludes costs of government restructuring and privatization revenues. The cyclically neutral level of revenue is defined to be a constant level of revenue relative to GDP. The cyclically neutral level of expenditure is defined to be a constant level of expenditure relative to potential output. The fiscal stance is the difference between cyclically neutral balance and actual balance. The fiscal impulse is the change in the fiscal stance from base year.

Source: World Bank staff estimates based upon IMF.

present macroeconomic circumstances. The decision to adopt a more expansionary fiscal stance in the future must be judged against a few key principles:

- Any increase in the fiscal deficit must be financed through viable borrowing rather than money creation. Rising prices hurt the poor disproportionately, and low-income groups could well lose more through the inflation tax than they gain through new employment opportunities.

- Financing the increased fiscal deficit should avoid upward pressure on domestic interest rates and minimize the borrowing costs of the whole economy. While in the short term, most countries can borrow locally and absorb liquidy without creating upward pressure on interest rates, as recovery takes hold, government may borrow locally only by crowding out potential private investment and pushing up interest rates, which jeopardizes the fragile corporate sector. This suggests that governments should seek to borrow under their sovereign status in international markets or through international financial institutions.[4]

- Efficiency improvements to public spending should accompany expanded deficit spending. Increasing spending in government enterprises with unsold surplus products—a problem in China—or other inefficient spending will not ultimately pay for the debt that has financed them. Governments throughout the region have ample scope to increase their expenditure efficiency through better management and better project evaluation data to feed back into the budget process.

If fiscal policy were to become more expansionary, increased net spending must support the overall objectives of generating a rapid, broadly shared recovery. Spending increases vary from country to country, but they should focus on the following:

- Support for low-income groups is a priority: employment programs to construct labor intensive public works, such as rural roads, and environmental projects, such as reforestation, sanitary, and water facilities; insurance or transitional income support for the unemployed; increased subsidies for basic foods; and health outlays. Expenditures that result in rising incomes for the poor will, because of their higher propensity to consume, have an added direct effect on aggregate demand.

- Fiscal measures for the corporate sector can help investment, restructuring, and exports: these might include time-bound tax incentives for investment; temporary exemption for income and capital gains taxes; and tax rebates for exports and reduced tariff rates.

- Tax reductions have the advantage of putting resources in the hands of would-be consumers, but the disadvantage of not being targeted to redress inequity or to reactivate the economy. Since the poor pay fewer taxes, tax reductions tend to benefit the relatively well off. Also, in East Asia, people may well prefer to save during times of crisis, and this could dilute the effect of tax cuts.

Using some public funds for bank recapitalization is unlikely to stimulate aggregate demand directly since these resources go into bank capital rather than spending (and for that reason have been left out of the fiscal stimulus analysis above). While public funds may be justified if it can be shown that faster recapitalization will boost intermediation by the financial sector, and there are complementary measures that allocate losses fairly among owners and creditors. Moreover, fiscal expenditure for banking recapitalization is fundamentally different from other "normal" fiscal expenditures to the extent that it recognizes contingent liabilities. Also, annual fiscal costs and government debts can be reduced once the recapitalized banks are sold to the private sector in the near to medium term. This is not automatic, however, and history is replete with examples of social losses that remain a burden to taxpayers for years.

Because net debt levels among East Asian governments (except for the Philippines) prior to the crisis were relatively low, government generally can manage additional indebtedness. This capacity is not infinite: governments have suffered a capital loss associated with the exchange rate depreciation, and they also have to shoulder the burdens of the fiscal workouts. Nonetheless, the borrowing costs associated with a more expansionary fiscal position are likely to be manageable.

Concerted fiscal stimulus

If fiscal stimulus were undertaken simultaneously, the increase in demand would produce a strong impulse that could counteract the recessionary forces in the region—especially if Japan were to take a strong lead. A parallel, concerted policy of fiscal of 1 percent GDP increase in deficit-spending on the main countries of the region—including Japan, Asian NIEs, ASEAN 4, and China—would have a major impact. If these actions were undertaken together in the third and fourth quarter of 1998, they could generate new demand that could lift East Asia by 2.0 percentage points of GDP.[6] This policy would also have a positive effect on U.S. and EU economic growth as well, perhaps by as much as 0.3 percent of GDP.

In practice, levels of increases in deficit spending must be tailored to each country. Some economies—Taiwan (China) and Singapore, perhaps—may be operating at lower rates of capacity utilization and have low debt levels, and therefore might have some scope for increases in spending. Other countries—the heavily indebted Philippines—might opt for lower increases. Nonetheless, concerted fiscal policy action tends to countervail the enormous drag that the economies of the region are having on one another.

Financing these new expenditure levels will have to take care that they do not have adverse impact on domestic interest rates. Economies with plentiful internal savings and well developed capital markets—including Japan, Singapore, and Taiwan (China)—can finance these amounts in their domestic markets. China could as well. Additional domestic borrowing in these countries would not crowd out private investment, nor will the effects on interest rates dampen new private investment. The remaining countries would do better to seek long-term funds from abroad. The amount to finance this level of stimulus would be less than US$10 billion.

Monetary and interest rate policy

Interest rates have come down in recent months—except in Indonesia—to new, post-crisis lows. Nominal interest rates are now hovering near their pre-crisis levels in Korea, Thailand, and Malaysia, and real interest rates are now below those prevailing in June 1997 in these countries and the Philippines. This is welcome news, especially in light of the continued stability of the exchange rates. However, interest costs remain much larger because of a debt stock that has been made much larger by the effective capitalization of interest over the last year and the currency depreciation. As a result, the interest service burden remains heavy.

Since currencies in several countries appear to be stabilizing, there is no need to maintain a monetary policy that tightens liquidity conditions in East Asia. Interest rates should be allowed to fall further, consistent with maintaining exchange rate stability and maintaining the attractiveness of deposits in the financial system. Even under a more accommodating monetary policy lending, interest rates are likely to remain high until the effects of inflation wind their way through the system, until investment risk premia abate, and until banks recapitalize. To achieve sustainable interest rates declines, a more accommodating monetary policy will need to be part of continued program of structural reforms in the financial and corporate sectors, and be made in the context of assurances about adequate external financing.

Lower interest rates may help improve creditworthiness and reduce debt-servicing costs, but they are likely insufficient to restart the flow of credit in affected economies. Such credit is needed desperately not only to resume normal business operations but also to facilitate the substantial resource shifts associated with structural reforms, including shifts from the dramatic rate realignment. Export credit is an obvious priority, but also there may be high returns to special credit for small- and medium-sized businesses and for small farmers in rural areas, segments of the financial market that are suffering a disproportionate impact of the banking crisis and whose employment impacts are great.

Protecting the poor and sharing a recovery

The crisis has pushed millions into the abyss of poverty. Rekindling economic growth is the only way out. A growing economy creates jobs and income. Public actions can make an important difference to the fate of the poor, both in mitigating the effects of recession and in ensuring that recovery is broadly shared. The immediate agenda includes:

- *Pro-poor macroeconomic policies.* A fiscal stimulus directed at labor-intensive activities, such as rural roads, rural environmental projects, sanitation, and some construction projects, would combine the benefits of growth with benefits for low-income groups.

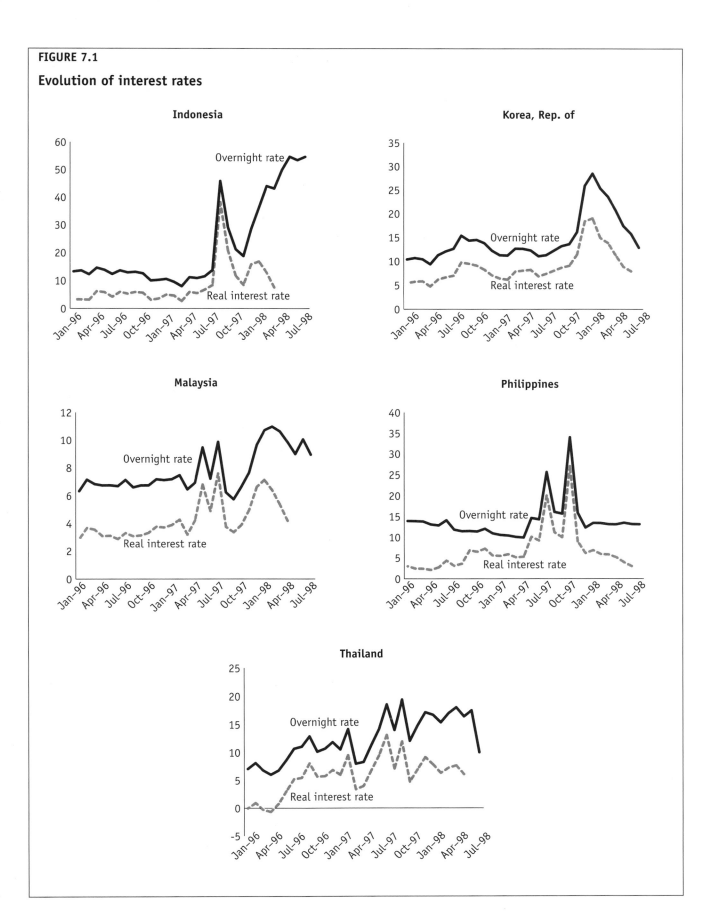

FIGURE 7.1

Evolution of interest rates

Indonesia

Overnight rate

Real interest rate

Korea, Rep. of

Overnight rate

Real interest rate

Malaysia

Overnight rate

Real interest rate

Philippines

Overnight rate

Real interest rate

Thailand

Overnight rate

Real interest rate

- *Protecting the consumption of the poor.* This could be accomplished by ensuring access to food and preserving the purchasing power of vulnerable households. Special efforts should be made to ensure that women receive benefits directly since they and their children are most adversely affected. For the unemployed, a system of unemployment assistance benefits could provide for income supplements for a limited time in the form of flat benefits at around the poverty line. Well-targeted public employment programs are an alternative.
- *Ensuring the poor have access to public services.* During the crisis, the poor stand to suffer from irreversible losses in potential education and health that will impede their participation in future recovery. Efforts to keep schools and health care affordable for poor households and quality of services intact are particularly important. This will undoubtedly entail preserving real spending levels on primary schools, and seeking to maintain nonsalary spending levels. Increasing targeted subsidies to encourage students, especially girls, to stay in secondary school, linked as closely as possible to income level, would also help.

When recovery does take hold, governments will have to return to the precrisis agenda of long-term challenges. They will have to do so with a new urgency. Insecurity has increased as a consequence of the crisis. Mechanisms are needed to help households better manage insecurity related to old age and health and employment risks. Labor market policies that permit collective bargaining without introducing rigidities or labor monopolies are needed, together with a larger role for the private sector in providing job training and job placement services. Education policies that expand access to secondary education and upgrades labor skills can mitigate market pressures toward inequality and contribute to increasing national competitiveness. Not only will these future reforms help protect low-income groups as recovery takes hold, they can form part of the foundation of more stable economies. Funded pension systems, for example, provide the institutional investors necessary to supervise corporate performance at the same time they provide greater stability for domestic bond and equity markets.

Progress on structural reforms: Improving the quality of growth

Structural reforms are necessary but insufficient for immediate economic recovery. Nevertheless they are vital to lay the foundation for sustained growth. Moreover, when reforms begin taking effect, investor confidence is boosted so that the economies can take advantage of macroeconomic stability and incipient demand reactivation to capture new sources of risk capital and launch a recovery.

The steady torrent of revised downward projections of the region's prospects has drowned out news of the enormous progress on structural reforms. This trend is true not only for the East Asia 5, but also China, and some of the smaller countries. All of the crisis-affected countries have improved legislation affecting their financial sectors—supervision, prudential regulation, transparency, and conditions of entry—especially from foreign competitors. Governments must continue to progress on more profound institutional reform agendas in the financial sector, including:
- Phase out heavy state involvement in credit and systems control
- Strengthen the regulatory framework by adopting supervision, loan classification standards and loan provisioning in line with international practices, plus progressively raising capital adequacy requirements and increasing transparency and disclosure for financial institutions
- Transform banks' governance by creating roles for outside directors and investors, and clarifying legal responsibilities for managers of financial institutions
- Reduce restrictions on foreign ownership; and
- Improve the legal framework for resolving distressed financial institutions and nonperforming assets.

The *corporate sectors*, formerly insulated from the threat of takeover or bankruptcy, are now exposed legally to both. Developing corporate bond markets heralds more balanced financial systems that will rely less on bank financing, diversify risk, and improve corporate monitoring. New laws improve competition, more clearly align interests of management and shareholders through use of outside directors, protect minority shareholder interests, and improve public reporting and standard disclosure. All countries have recently tried to facilitate enterprise restructuring through

BOX 7.4

It takes a long time to regain investors' confidence...

Agencies are often slow to catch problems affecting sovereign and private commercial paper. But, they are also slow to certify renewed creditworthiness. Both points are brought out in examining ratings of Indonesia, Korea, and Thailand, with Finland and Sweden, countries similarly afflicted with banking crises in 1991. If the ratings for Finland and Sweden are any indication, it will take several years for countries to regain precrisis sovereign ratings. In the case of bank/corporate ratings, it takes even longer as these countries are still well below their precrisis ratings. This implies that banks and corporations in crisis-stricken countries will have to endure higher borrowing costs in years to come.

Source: World Bank staff and Moody's Investors Service.

Ratings of sovereign, banks, and corporations

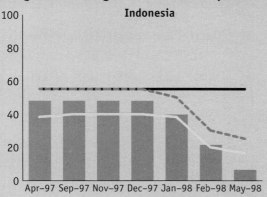

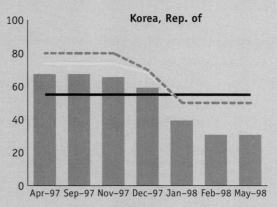

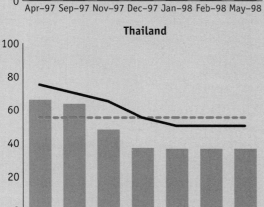

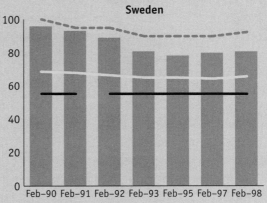

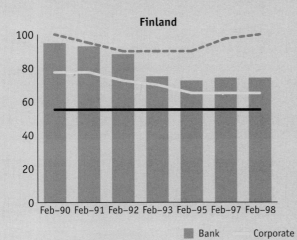

AAA = 100	Ba1 = 50	C = 0
Aa1 = 95	Ba2 = 45	
Aa2 = 90	Ba3 = 40	
Aa3 = 85	B1 = 35	
A1 = 80	B2 = 30	
A2 = 75	B3 = 25	
A3 = 70	Caa1 = 20	
Baa1 = 65	Caa2 = 15	
Baa2 = 60	Caa3 = 10	
Baa3 =55	Ca = 5	

Bank   Corporate   -- Sovereign   ━ Investment grade

BOX 7.5

Why did one tiger escape barely wounded?

The economy of Taiwan, China showed none of the macroeconomic or structural imbalances that contributed to the regional financial crisis. Yet its currency and stock market were affected by contagion, while export and growth prospects are down. Taiwan, China ranks among the larger economies in Asia; it has a per capita income of about US$12,000 and a population of about 22 million. On average, GDP grew by 6.4 percent from 1990 to 1997 (6.8 percent in 1997). The current account has been typically in surplus (2.7 percent of GDP in 1997). Public external debt is negligible at US$100 million. Private external debt stood at US$7.5 billion at the end of 1997, about 9 percent of official foreign exchange reserves. The capital account was gradually liberalized and there are relatively few restrictions in place at present. The year 2000 was set as the target for complete liberalization. The budget deficit in fiscal 1997 was 2.3 percent of GDP and public domestic debt at the end of 1997 stood at 24 percent of GDP. The banks are generally well capitalized. Their foreign currency liabilities are fully matched by foreign currency assets and there are no systemic maturity mismatches. They have low exposure to the stock market and to real estate. Their overdue loan ratio stood at less than 4 percent at the end of 1997. The industrial sector is dominated by SMEs with an exceptionally modest average debt/equity ratio of 0.7 at the end of 1997 (about one-seventh the corresponding average ratio for Korean's chaebols). Well functioning bankruptcy procedures permit the relatively easy exit of insolvent firms.

Despite exceptionally strong macroeconomic and structural indicators, contagion led to a speculative attack. In defense of the exchange rate, interest rates were allowed to rise while the central bank intervened. Between August and mid-October 1997, US$7 billion in reserves were lost, 1997—US$3 billion alone in the week preceding October 17, 1997, when officials decided to float the rate. The following day a major speculative attack on the Hong Kong, China dollar triggered an unprecedented 29.4 percent drop of the Hang Seng index. The currency of Taiwan, China depreciated by 15 percent to NT$32.6 per U.S. dollar by the end of 1997. The rate stabilized during the first four months of 1998, but fell to NT$34.5 per U.S. dollar in June in response to the continuing slide of the Japanese yen and instability in Indonesia. Contagion also affected the stock market in Taiwan, China. The local index continued to rise during the first 8 months of 1997, but then fell by 30 percent in the following 6 weeks, contributing to the fear that continuing regional financial turmoil would lead to intensified pressure on the currency. Equity prices partially recovered since the floating of the currency on October 17, 1997, but from early March 1998, the trend has been generally down due to Japan's worsening recession. During the first half of 1998, exports fell by 7 percent (yoy); industrial output growth decelerated and inventories accumulated. GDP growth in 1998 is expected to fall slightly to 6 percent or below.
Source: World Bank staff.

improving the enabling environment, including better accounting and disclosure standards, enhanced bankruptcy and foreclosure processes, and changes in taxation and accounting rules.

A main element of such reforms is to introduce truly transparent accounting and auditing systems, consistent with international best practice. This would require (a) reducing the role of the government in regulating and overseeing the accounting and auditing practices and profession; (b) establishing an independent and self-regulating national professional body for setting accounting standards; and (c) strengthening the financial oversight functions of boards of directors and improving the effectiveness of audit listed companies by establishing audit committees of boards of directors. Crisis countries are taking several steps in the direction of these reforms. In Thailand, the government is committed to giving legal authority for accounting standard-setting to an independent organization, the Thailand Accounting Standard Board, which will con-

sist of members from the accounting profession, business, government, and academia. The Institute of Certified Accountants and Auditors of Thailand is also set to review the standards and regulatory framework of the profession and to propose amendments to applicable laws. Similar measures are being taken in Korea. Foreclosure of collateral remains generally weak; in many East Asian countries it often took several years. This will take considerable effort to improve.

China confronts unique internal challenges that echo strongly those elsewhere in the region: the need to strengthen the banking system, improve prudential supervision, and reduce political interference in lending decisions. Reforming its 300,000 state enterprises, with their 100 million workers, requires a restructuring process that is more vast in dimension and complicated in execution than perhaps anywhere else in the world. But in China, too, progress has been formidable.[7] Vietnam confronts similar problems. Neither country can afford to delay these reforms.

BOX 7.6

Learning from the Asian crisis

Though it is still early days, lessons are emerging from East Asia, both about the prevention of crisis and the response to it. Preventing crisis requires:

- *Reducing the incentives for excessive borrowing.* Thailand, Korea, and Indonesia created policy-induced incentives to borrow through implicit and explicit guarantees, exchange rate policies, and lax rules governing public disclosure and reporting. Removing these incentives would mitigate inflow-fueled credit booms.
- *Improving governance in the financial and corporate sectors.* Countries with weak capacity to intermediate credit, especially from abroad, and with weak systems of corporate governance experienced the most severe problems. Reducing the government-directed credits, increasing capital adequacy ratios, and strengthening regulation of financial intermediation as well as improving bankruptcy laws, accounting and auditing, and disclosure requirements can play important roles in ensuring that capital will be well invested.
- *Enhancing prudential regulation, especially for short-term capital flows.* In crisis, the distinction between private and public debt easily blurs. Private external bank debt tends to become socialized to prevent collapse of the banking system, and thus constitutes huge and implicit contingent liabilities. To prevent an excessive build-up of short-term liabilities, governments should monitor short-term debt and establish tight prudential regulations limiting exposure of financial institutions. Similarly, better disclosure of information would reduce one source of speculation affecting investors' outlook.

Even the best of policies cannot guarantee that financial crises will not occur. Many well-regulated and supervised developed countries have had crises in the last decade. However, good regulation can reduce the probability of crises and minimize their costs.

Responding to crises better requires, among other things:

- *Improving bankruptcy laws and mechanisms for restructuring.* The absence of credible and well established bankruptcy provisions not only reduced the incentive for corporate managers to allocate capital to its best use before the crisis, but has hobbled the restructuring process once companies became insolvent. Without credible laws and procedures, banks and

corporate owners may delay recognizing their losses to prevent losing control of assets to creditors or new investors, and the financial and operational restructuring could drag on for years. As the crisis has become systemic, the scale of the problem has shown the need for establishing mechanisms to encourage rapid and voluntary mechanisms for creditors and potentially viable companies to restructure as an alternative to court-ordered bankruptcy and liquidation of assets.

- *Enhancing social protection.* Establishing effective systems to protect the socially vulnerable and allocate public resources for anti-poverty programs allows countries to better target resources to groups of people who are hardest hit by the crisis. This includes systems of unemployment insurance, pension, health protection, and income maintenance for low-income groups.

Other questions are now the subject of debate. Two important ones under study are:

- *Does the international system require a new architecture to avoid volatility, contagion, and increased moral hazard?* The Asian crisis has underscored the need for better mechanisms for facilitating debt workouts, especially private-to-private. Under exceptional circumstances, new mechanisms, perhaps including possible debt standstills and revisions to domestic bankruptcy laws to take into account large unusual movements in the exchange rate, may be necessary to prevent funding crises from becoming systemic and to accelerate private-to-private debt resolutions. Similarly, discussions are underway to consider whether mechanisms could be established to mitigate disastrous financial panics without creating "moral hazard" incentives to over-lend.
- Are there alternative macroeconomic responses to crises that might establish the conditions of growth more quickly than those adopted in the first days of the East Asian crisis? With hindsight, could policy makers have adopted a different combination of policies that would have led to quicker reactivation and restoration of investor confidence? The question is not easy because calibrating programs requires, among other things, judgments about the responsiveness of private agents to policies.

Investments in people and in the institutions governing human resources are transforming labor markets. In the short run, labor rules that protected established workers more than new entrants have been changed so that the inevitable needs for greater flexibility are combined with unemployment insurance, worker compensation, and new training opportunities. In the medium term, countries are seeking to increase investments in education and health, plus protect the elderly, through newly funded pension programs.

The *environment* has almost disappeared from the agenda since the crisis. Ironically, by shrinking people's incomes, the crisis has forced conservation on the entire region. Lower demand has cut energy-related air and water pollution, and reduced aggregate pressure on natural resources (even though pressures in isolated areas may have increased, for example, as newly impoverished people are forced to substitute wood fuels for commercially purchased fuels). But, forced conservation is neither desireable nor sustainable. In striving for recovery, the policies that countries adopt today to generate growth will shape the environment of tomorrow. Passing through energy price increases associated with exchange rate descrepancies, adopting policies to prevent pollution (such as pollution taxes and price adjustments) rather than focusing on end-of-pipe controls, and investing in sanitation and clean water, perhaps as part of a fiscal stimulus program, can have long-term returns.

Greater consultation among countries in the region on policy architecture and implementation holds considerable promise for improving the quality of policies and raising their probability of success. If macroeconomic policies could be coordinated, if policies toward foreign competition and corporate governance issues could attain some degree of harmony, and if more cooperation led to greater progress in reducing trade barriers, the region as a whole would benefit. Similarly, simply sharing experiences on particular policies—toward private financing of infrastructure and in reducing corruption—can speed dissemination of learning about best practices in these areas. Various mechanisms now exist to coordinate regional policies. Fora provided through the Asia Pacific Economic Cooperation (APEC), the recently established Manila Framework, Association of South East Asia Nations (ASEAN), regular meetings of the governors of central banks in the Executive Meetings of East Asia and Pacific (EMEAP), as well as international meetings associated with the World Bank and IMF provide abundant opportunity that governments should increasingly seize.

By all indications, the region is enacting one of the most remarkable structural transformations in its history. In a couple of years, government's role in allocating financial resources will decline, but its responsibility for supervising that process will increase. Business deals will likely be based less on personal relationships between elite entrepreneurs and managers and their bankers and more on impersonal bond and equity markets. Ownership is likely to be much more diffuse. Societies will rely much less on kinship ties and rapid growth to protect their sick, elderly, and unemployed, and much more on new formal institutions. Even if these policies are secondary to the immediate agenda of reactivating the economy, they will help restore confidence in economic management and position the region to sustain a recovery once it has begun.

Mobilizing additional resources to finance growth

Capital flows are the lifeblood of economic growth, and the region's capital hemorrhage must be staunched and reversed. Restoration of capital flows is a high priority. This would help relieve the compression of private consumption that comes with the massive swing to a current account surplus. Restoring capital flows can only be done by reactivating the economies and providing a context that will restore investor confidence.

The countries are doing their part to restore confidence. They have enacted difficult structural reforms, opened their markets wider to foreign direct investment, and protected the creditworthiness of their economies by assuming financial system liabilities. But, the large pools of investment capital that deserted East Asia are still waiting for some signs of economic reactivation.

The partnership of international financial institutions, governments, and private banks should explore ways to mobilize new capital. Financing a 1 percent of GDP stimulus in Thailand, Indonesia, Malaysia, and Korea would require that the international community mobilize US$10 billion. Compared to the US$100 billion swing in capital flows against the region, this amount seems relatively small. As governments begin to restructure their financial systems and negotiate with external lenders, the international financial institutions should work with the commercial banks to "bail them in" on a proportional basis while they also mobilize international additional capital.

While virtually all capital is welcome under the circumstances of withdrawal of short-term capital, the region particularly needs long-term portfolio and foreign direct investment. Governments in the region will have to think carefully about the regime that integrates their domestic financial sectors with foreign capital flows.[8]

The journey ahead

East Asia's crisis is still unfolding and its full course is not yet clear. Nonetheless, the policy challenges are becoming clear: restructuring the banking and corporate sector, reactivating now-depressed domestic demand, and adopting policies that will protect the poor in the difficult year ahead, but that will also include them centrally in the recovery that will eventually come. The response of the international community is particularly important. Only by restoring capital flows can East Asia resume growth, and only by resuming growth can it reverse the massive income losses that have imperiled the livelihoods of so many people in such a short time.

Notes

1. Consensus Economics, *Asia Pacific Consensus Forecasts,* August 1998.

2. Systemic Bank Restructuring and Macroeconomic Policy (1997), (eds.) W. Alexander, J. M. Davis, L, Ebrill, and L. Lindergren.

3. It should be noted that the actual fiscal balance for a given period may be different than the targeted level to the degree that agreed policies are not undertaken or economic conditions differ. In Thailand, for example, the actual fiscal balance was in deficit during the first quarter of 1997–98, notwithstanding the early fiscal objective of aiming for a small surplus for the year as a whole.

4. To be sure, these calculations are somewhat sensitive to assumptions about "full capacity" (that is, potential output); in some countries, this potential may have shrunk because at least some of the installed capacity was probably wasted in industries or activities where the country would not be competitive.

5. The international financial institutions face severe constraints under present circumstances. The World Bank, for example, is constrained by the need to maintain its own strict capital-lending ratio.

6. This multiplier is predicated upon an analysis of cross-country demand effects through trade linkages among countries, and is obtained for the East Asia economies as a whole, excluding Japan.

7. See World Bank *China 2020: Development Challenges in the New Century,* September 1997. For an update on recent policies toward state enterprises, see Richard Newfarmer and Dana Liu "China's Race with Globalization" *China Business Review,* July-August 1998 25:4.

8. For a thorough analysis of policies affecting capital flows, see World Bank, *Managing Capital Flows in East Asia*, 1996.

References

Chapter 1

Alba, P. A. Bhattacharya, S. Claessens, S. Ghosh, and L. Hernandez "Volatility and Contagion in a Financially-Integrated World: Lessons from East Asia's Recent Experience" Paper presented to the PAFTAD 24 conference "Asia Pacific Financial Liberalization and Reform, May 20–22, 1998, Chiangmai, Thaland.

Asian Development Bank. *Emerging Asia*. 1997.

Bosworth, B. and S. Collins. 1996. "Economic Growth in East Asia — Accumulation Versus Assimilation." Brookings Papers on Economic Activity 2, 135–204.

Hsieh, C. 1997. "What Explains the Industrial Revolution in East Asia? Evidence from Factor Markets." Manuscript, UC Berkeley.

Kaminsky, Graciela, and Carmen Reinhard. 1997. "The Twin Crises: The Causes of Banking and Balance of Payments Problems."

Kim, J. and L. Lau. 1994. "The Sources of Growth of the East Asian Newly-Industrialized Countries." *Journal of the Japanese and International Economies* 8(3), 235–71.

Krugman, Paul. 1994. "The Myth of Asia's Miracle". *Foreign Affairs* 73 (6), 62–78.

McKibbon, Warwick. 1998. "The Crisis in Asia: an Empirical Assessment."

Nehru, V. and A. Dareshwar. 1993. "A New Database on Physical Capital Stock: Sources, Methodology and Results." Revista de Analisis Economico 8 (1), 37–59.

Rodrik, D. 1997. "TFPG Controversies, Institutions and Economic Performance in East Asia." National Bureau of Economic Research Working Paper No. 1587.

Sarel, M. 1997. "Growth and Productivity in ASEAN Countries." IMF Working Paper No. 97/97.

"The East Asian Miracle: Ecomonic growth and public policy." New York, NY: Published for the World Bank (by) Oxford University Press, 1993.

World Bank. *Global Economic Prospects 1998*. Draft.

Young, A. 1992. "A Tale of Two Cities: Factor Accumulation and Technical Change in Hong Kong and Singapore," in O. Blanchard and S. Fischer, eds., NBER *Macroeconomics Annual*. Cambridge: MIT Press.

Young, A. 1994. "Lessons from the East Asian NICs: A Contrarian View." *European Economic Review* 38 (3–4), 964–73.

Young, A. 1995. "The Tyranny of Numbers: Confronting the Statistical Realities of the East Asian Growth Experience." *Quarterly Journal of Economics* 110 (3), 641–80.

Chapter 2

Andersen, Lykke, Amit Dar, Martin Godfrey, Chris Manning, Dipak Mazumdar and Zafiris Tzannatos. 1997. "Growth and Inequality in Thailand: An Overview of Labor Market Issues." Mimeo. December 5.

Bhattacharya, Amar, Swati Ghosh and Jansen. 1997. "Note on East Asia's Export Performance." mimeo. World Bank.

Dasgupta, Dipak, and Kumiko Imai. 1997. "The 1996 Slowdown in East Asia's Exports: Structural or Cyclical Factors?" Conference Proceedings, AEA Eastern Association.

Dasgupta, Dipak, Bejoy Dasgupta and Edison Hulu. 1995. "The Determinants of Indonesia's Non-Oil Exports." Conference Paper on Deregulation. EDI/World Bank.

Dasgupta, Bejoy. 1989. "Exports and Exchange Rate Policy: The Case of India." D.Phil Thesis, University of Oxford.

Diwan, Ishac and Bernard Hoekman. 1998. "Competition, Complementarity and Contagion in East Asia." CEPR/EDI Conference Paper on Financial Crisis. World Bank.

Dollar, David and Mary Hallward-Driemeier. 1998. "Crisis, Adjustment and Reform: Results from the Thailand Industrial Survey." Mimeo. World Bank.

Ernst, Dietr. 1998. "Destroying or Upgrading the engine of growth? The reshaping of the electronics industry in East Asia after the Crisis." Background paper for this study. Mimeo. World Bank.

Faini, Riccardo, Fernando Clavijo, and Abdel Senhadji-Semlali. 1990. "The Fallacy of Composition Argument: Does Demand Matter for LDC Manufactured Exports?" Center for Economic Policy Research Discussion Paper No. 499S.

Feenstra, Robert. 1994. "New Product Varieties and the Measurement of International Prices." Amercian Economic Review 84 (1), April 157–177.

Feenstra, Robert, and Andrew K. Rose. 1997. "Putting Things in Order: Patterns of Trade Dynamics and Growth." NBER Working Paper No. 5975.

Goldberg, P.K., and M.M. Knetter. 1995. "Measuring the Intensity of Competition in Export Markets." NBER Working Paper No. 5226

Goldstein, Morris, and Mohsin Khan. 1985. "Income and Price Effects in Foreign Trade." Ronald Jones and Peter B. Kenen, ed. *Handbook of International Economics*, Vol. 2. Amsterdam and New York: North-Holland and Elsevier.

Grossman, Gene M. 1982. "Import Competition From Developed and Developing Countries." *The Review of Economics and Statistics*, 64 (2), May. 271–281.

Grubel, H.G. and P.J. Lloyd. 1975. *Intra-Industry Trade: The Theory and Measurement of International Trade in Differentiated Products*. New York: Halstead Press.

Helpman, Elehanan. 1986. "Imperfect Competition and International Trade: Evidence From Fourteen Industrial Countries." *Journal of the Japanese and International Economics*, Vol. 1. 62–81.

Hoekman, Bernard, and Simeon Djankov. 1996. "Intra-Industry Trade, Foreign Direct Investment, and the Reorientation of East European Exports." World Bank Policy Research Working Paper No. 1652, September.

Kawai, Masahiro. 1997. "Japan's Trade and Investment in East Asia." In David Robertson, ed., *East Asian Trade after the Uruguay Round*. Cambridge: Cambridge University Press, 209–226. [A more detailed version: Kawai, Masahiro. "Interactions of Japan's Trade and Investment: A Special Emphasis on East Asia." University of Tokyo Discussion Paper Series No. F-39, October 1994.]

Kawai, Masahiro. 1997. "East Asia Currency Turbulence: Implications of Financial System Fragility." Mimeo. World Bank.

Kawai, Masahiro. 1998. "East Asian Currency Crisis: Causes and Lessons." *Contemporary Economic Policy*, vol. 16, April, 157–172.

Kawai, Masahiro, and Shujiro Urata. 1996. "Trade Imbalances and Japanese Foreign Direct Investment: Bilateral and Triangular Issues." In Ku-Hyun Jung and Jang-Hee Yoo, eds., *Asia Pacific Economic Cooperation: Current Issues and Agenda for the Future*, East and West Studies Series, 39, Institute of East and West Studies, Yonsei University, October, 61–87.

Kawai, Masahiro, and Shujiro Urata. 1998. "Are Trade and Investment Substitutes or Complements? An Empirical Analysis of Japanese Manufacturing Industries." In Hiro Lee and David W. Roland-Holst, eds., *Economic Development and Cooperation in the Pacific Basin: Trade,*

Investment and Environmental Issues. Cambridge: Cambridge University Press.

Kishimoto, T. 1998. "Korea's Semiconductor Industry and the Currency Crisis." Nomura Research Institute. Draft.

Noland, Marcus. 1997. "Has Asian Export Performance Been Unique?" *Journal of International Economics*, Vol. 43. 79–101.

Lee, Jan. 1997. "Changing Trade Patterns in Asia." HSBC.

Li, Yangyang. 1998. "China's Exchange Rate Policy." Mimeo. World Bank.

Lipsey, Robert E. 1994. "Quality Change and Other Influences on Measures of Export Prices of Manufactured Goods and Primary Products and Manufactures." NBER Working Paper No. 4671.

Muscatelli, V.A., and C. Montagna. 1994. "Intra-NIE Competition in Exports of Manufactures." *Journal of International Economics*, Vol. 37. 29–47.

Chapter 3

Akerlof, George and Paul Romer. 1993. "Looting the Economic Underworld of Bankruptcy for Profit." Brookings Papers on Economic Activity, 2:1–73,1993.

Asian Development Bank and World Bank. 1998. "Managing Global Financial Integration In Asia: Emerging Lessons and Prospective Challenges," March 10–12, 1998

Bernanke, Ben S. and Alan S. Blinder. 1988. "Credit, money, and aggregate demand." *American Economic Review*, Papers and Proceedings, 78:435–45 May.

Bhattacharya, Amar, Swati Ghosh and Jos Jansen. 1998. "Has the Emergence of China Hurt Asian Exports?" Mimeo. World Bank.

Blanchard, Olivier and Mark Watson. 1982. "Bubbles, Rational Expectations and Financial Markets." Paul Wachtel, ed., *Crises in the Economic and Financial Structure* (Lexington Books).

Calvo, Sara, and Carmen Reinhart. 1995. "Capital Flows to Latin America: Is There Evidence of Contagion Effects?" Unpublished manuscript. The World Bank—International Monetary Fund.

Claessens, Stijn and Thomas Glaessner. 1997. "Are Financial Sector Weaknesses Undermining the East Asian Miracle." Directions in Development. World Bank. September.

Claessens, Stijn and Thomas Glaessner. 1998. "Internationalization of Financial Services in Asia." Working paper, No. 1911, World Bank.

Corsetti, Giancarlo, Paolo Pesenti and Nouriel Roubini. 1998. "What Caused the Asian Currency and Financial Crisis?" Mimeo. New York University.

Demirgüç-Kunt, Asli, and E. Detragiache. 1998. "Financial Liberalization and Financial Fragility." World Bank Annual Bank Conference on Development Economics Paper. April 20–21, 1998. Washington DC.

Demirgüç-Kunt, Asli, and H. Huizinga. 1998. "Determinants of Commercial Bank Interest Margins and Refutability: Some International Evidence." Policy Research Working Paper Series No. 1900. World Bank. May.

Diamond, Douglas W., and Philip H. Dybvig. 1983. "Bank runs, deposit insurance, and liquidity," *Journal of Political Economy*, 91:401–19 June.

Ding, Wei, Ilker Domaç and Giovanni Ferri. 1998. "Is there a Credit Crunch in East Asia?" Policy Research Working Paper Series. World Bank, June.

Domaç, Ilker and Giovanni Ferri. 1998. "The Real Impact of Financial Shocks: Evidence from Korea." Mimeo. World Bank.

Eichengreen, Barry, and Ashoka Mody. 1998. "What Explains Changing Spreads on Emerging-Market Debt: Fundamentals or Market Sentiment?" mimeo. IMF and World Bank, paper prepared for NBER-conference *Capital Inflows to Emerging Markets.*

Feldstein, Martin. 1998. "Refocusing the IMF." *Foreign Affairs*, 77:20–33, March/April 1998.

International Monetary Fund. 1997. *World Economic Outlook.* December. Washington, DC.

Kaminsky, Graciela, and Carmen Reinhart. 1996. "The Twin Crises: The Causes of Banking and Balance of Payments Problems." Unpublished paper, Federal Reserve Board.

Kaminsky, Graciela and Sergio Schmukler. 1998. "On Booms and Crashes: Is Asia Different?" Mimeo. World Bank and Board of Governors of the Federal Reserve System.

Kaufman, Daniel, Gil Mehrez, and Sergio Schmukler. 1998. "The East Asian Crisis: Was It Expected?" Mimeo. World Bank.

Kawai, Masahiro and Kentaro Iwatsubo. 1998. "The Thai Financial System and the Baht Crisis: Processes, Causes and Lessons." Mimeo. Institute of Social Science, University of Tokyo.

Krugman, Paul. 1979. "A Model of Balance of Payments Crises." *Journal of Money, Credit and Banking*, IV. 11. pp. 311–325.

_____. 1998. "Fire-Sale FDI." mimeo. MIT, paper prepared for NBER-conference *Capital Inflows to Emerging Markets*.

La Porta, Rafael, Florencio Lopez-de-Silanes, Andrei Shleifer, and Robert W. Vishny. 1997. "Legal Determinants of External Finance." *Journal of Finance*, Vol. LII, Number 3, July.

La Porta, Rafael, Florencio Lopez-de-Silanes, Andrei Shleifer, and Robert W. Vishny. 1998. "Law and Finance." *Journal of Political Economy*.

Miller, Marcus, and Zhang. 1997. "Sovereign Liquidity Crises: The Strategic Case for a Payments Standstill." Mimeo. University of Warwick.

McKinnon, Ronald I. and Huw Pill. 1997. "Credible Liberalizations and International Capital Flows: The Over-Borrowing Syndrome." *American Economic Review*. Papers and Proceedings, 87. No. 2:189–93, May.

Montes, Manuel F., 1998. "The Currency Crisis in South East Asia." Published by Institute of South East Asian Studies.

Nomura Reserach Institute, 1998.

Obstfeld, Maurice. 1986. "Rational and Self-Fulfilling Balance of Payments Crises." *American Economic Review*, IV.76. pp. 72–81.

Obstfeld, Maurice. 1996. "Models of Currency Crises with Self-Fulfilling Features." *European Economic Review*.

Ramos, Roy. 1997. "Asian Banks at Risk: Solidity, Fragility." September, Banking Research. Goldman Sachs.

_____. 1997. "1998: Issues and Outlook: Cyclical Slowdowns, Structural Ills and the Odds for Recovery." December, Goldman Sachs: Banking Research.

Radelet, Steven and Jeffrey Sachs. 1998a. "The Onset of the East Asian Financial Crisis." (updated: March 30, 1998).

Radelet, Steven and Jeffrey Sachs. 1998b. "The East Asian Financial Crisis: Diagnosis, Remedies, Prospects." Brookings Papers on Economic Activity. Panel. Washington, D.C., March 26–27, 1998.

Sachs, Jeffrey. 1994a. "Russia's Struggle with Stabilization: Conceptual Issues and Evidence." in Michael Bruno and Boris Pleskovic, eds., *Proceedings of the Annual Conference on Development Economics*, 57–80. World Bank. Washington, DC.

Sachs, Jeffrey. 1994b. "Beyond Bretton Woods: A New Blueprint." In *The Economist* (U.K.); 333:23, 25, 27 October 1–7.

Sachs, Jeffrey, Aaron Tornell, and Andres Velasco. 1996. "Financial Crises in Emerging Markets: The Lessons from

1995." *Brookings Papers in Economic Activity*, pp.147–215.

Sarel, Michael. 1997. "Growth and Productivity in ASEAN Countries." IMF Working Paper, WP/97/97.

Stiglitz, Joseph and Marilou Uy. 1996. "Financial Markets, Public Policy and the East Asian Miracle." *The World Bank Research Observer* 11(2): 249–76.

Valdés, Rodrigo. 1996. "Emerging Markets Contagion: Evidence and Theory." Unpublished manuscript, Massachusetts Institute of Technology.

World Bank. 1997. *Private Capital Flows to Developing Countries*. Washington, DC.

World Bank. 1998. *Global Development Finance*. Washington, DC.

Chapter 4

Asian Development Bank and World Bank. 1998. "Managing Global Financial Integration In Asia: Emerging Lessons and Prospective Challenges." March 10–12, 1998.

Aoki, Masahiko and Hugh Patrick. 1994. *The Japanese main bank system: Its relevance for developing and transforming economies*—Oxford: Oxford University Press.

Allen, Franklin and Douglas Gale. 1995. "A welfare comparison of intermediaries and financial markets in Germany and the US." *European Economic Review*, 39:179–209.

Baird, Douglas. 1993. *The Elements of Bankruptcy*, Westbury Press, chapter 5.

Caprio, Jerry and Demirgüç-Kunt, Asli. 1997. "The Role of Long-Term Finance: Theory and Evidence." World Bank Working Paper 1746.

Claessens, Stijn, Simeon Djankov, and Giovanni Ferri. 1998. "Corporate Distress in East Asia: Assessing the Damage of Interest and Exchange Rate Shock." Mimeo. World Bank.

Demirgüç-Kunt, Asli, and Ross Levine. 1996. "Stock Market Development and Financial Intermediaries: Stylized Facts." *World Bank Economic Review*. 10:2, 291–321, Washington, DC.

Demirgüç-Kunt, Asli, and Vojislav Maksimovic. 1994. "Capital Structures in Developing Countries: Evidence from Ten Countries." World Bank Working Paper 1320.

Demirgüç-Kunt, Asli, and Vojislav Maksimovic. 1996. "Financial Constraints, Uses of Funds and Firm Growth: An International Comparison." World Bank Working Paper 1671.

Dollar, David, and Mary Hallward-Driemeier. 1998. "Crisis, Adjustment, and Reform: Results from the Thailand

Industrial Survey." Paper presented in the May 20–22, 1998, Competitiveness conference, Bangkok, Thailand.

Fan, Joseph and Larry Lang. 1998. "The Nature of Diversification." Mimeo. World Bank.

Fukao, Mitsuhiro. 1998. "Japanese Financial Instability and Weaknesses in the Corporate Governance Structure." Mimeo. Keio University.

Glen, Jack and Brian Pinto. 1994, "Debt or equity? How firms in developing countries choose." Discussion paper, International Finance Corporation, no. 22, Washington, DC.

Harris, Milton and Arthur Raviv. 1991. "Theory of capital structure." *Journal of Finance*, 46:297–355, March 1991.

Hoshi T., Kashyap A. and Scharfstein D. 1994. "Corporate Structure, Liquidity, and Investment: evidence from Japanese Industrial Groups." *Quarterly Journal of Economics*, vol. 6.

Hoshi T., Kashyap A. and Scharfstein D. 1990. "Bank Monitoring and Investment: Evidence from the Changing Structure of Japanese Corporate Banking Relationship." University of Chicago Press.

Jensen, Michael. 1986. "Agency Costs of Free Cash Flow, Corporate Finance and Takeovers." *American Economic Review*. Papers and Proceedings, 76:323–29 May.

Jensen, Michael and Meckling. 1976. "Theory of the Firm: Managerial Behavior, Agency Costs and Ownership Structure," *Journal of Financial Economics*, Vol. 3(4), 305–60.

Johnson, Bruce, Robert Magee, Nandu Nagarajan, and Harry Newman. 1985. "An Analysis of the Stock Price Reaction to Sudden Executive Deaths: Implications for the Management Labor Market." *Journal of Accounting and Economics*, 7: 151–174.

Khanna, Tarun and Palepu, Krishna. 1996. "Corporate Scope and (severe) Market Imperfections: An Empirical Analysis of Diversified Business Groups in an Emerging Economy." Graduate School of Business Administration, Harvard University, Boston, MA, March 1996.

Lang, Larry and Rene Stulz. 1994. "Tobin's q, Corporate Diversification and Firm Performance." *Journal of Political Economy*, 102, p. 1248–1280.

Lang, Larry and J. Doukas. 1998. "Corporate Performance, Direct Investments and International Diversification." Mimeo. World Bank.

La Porta, Rafael, Florencio Lopez-de-Silanes, Andrei Shleifer, and Robert W. Vishny. 1997. "Legal Determinants of External Finance." *Journal of Finance*, 52: 1131–1150.

La Porta, Rafael, Florencio Lopez-de-Silanes, Andrei Shleifer, and Robert W. Vishny. 1998. "Law and Finance." *Journal of Political Economy*, forthcoming.

Luders, Rolf. 1998. "Identification of, and Solution to, the Solvency Problem of the Financial Sector during the 1982–83 Crisis in Chile." World Bank mimeo.

Morck, Randall, Andrei Shleifer, and Robert W. Vishny. 1988. "Management Ownership and Market Valuation: An Empirical Analysis." *Journal of Financial Economics*, 20, 237–265.

OECD. 1998. *Corporate Governance: Improving Competitiveness and Access to Global Capital Markets.* April 2, Paris.

Posner, Richard A. 1998. "Creating a Legal Framework for Economic Development." *World Bank Research Observer*, 13:1, 1–11.

Price Waterhouse Management Consultants Ltd. 1997. "Corporate Governance in Thailand: A Price Waterhouse Study." Commissioned by the Stock Exchange of Thailand. January.

Prowse, Stephen. 1994. "Corporate governance in an international perspective: A survey of corporate control mechanisms among large firms in the United States, the United Kingdom, Japan and Germany." Bank for International Settlements, Economic papers, 1021–2515, no. 41.

Prowse, Stephen. 1998. "Corporate Governance in East Asia: A Framework for Analysis." Mimeo. World Bank.

Rajan, Raghuram, and Luigi Zingales. 1997. "The Firm as a Dedicated Hierarchy." Mimeo. University of Chicago.

Rajan, Raghuram, Henri Servaes, and Luigi Zingales. 1997. "The Costs of Diversity: The Diversification Discount and Inefficient Investment." mimeo. University of Chicago.

Saunders, Anthony, and Ingo Walter. 1994. *Universal Banking in the United States.* Oxford University Press: New York.

Scharfstein, David and Jeremy Stein. 1997. "The Dark Side of Internal Capital Markets: Divisional Rent-Seeking and Inefficient Investment." NBER working paper no 5969.

Scharfstein, David. 1997. "The Dark Side of Internal Capital Markets II: Evidence from Diversified Conglomerates." Mimeo. MIT.

Siam Business Information, Ltd. 1996. "Features of the Thai Private Sector and Characteristics of Thai Companies." Bangkok, Thailand.

Singh, Ajit, and Hamid, J. 1992. "Corporate financial patterns in developing countries." Technical Paper 1,

Washington, DC: World Bank and International Finance Corporation.

Singh, Ajit. 1995. "Corporate financial patterns in industrializing economies: A comparative study." Technical Paper 2, April, Washington, DC: World Bank and International Finance Corporation.

Shleifer, Andrei and Robert W. Vishny. 1997. "A Survey of Corporate Governance." *Journal of Finance*, 52: 737–83.

Stein, Jeremy. 1997. "Internal Capital Markets and the Competition for Corporate Resources." *Journal of Finance*, Vol. 52, pp. 111–134.

Stock Exchange of Thailand. 1997. "The roles, duties and responsibilities of the directors of listed companies." June 4.

United Kingdom. *Report of the Committee on the Financial Aspects of Corporate Governance*. Cadbury Report, 1 December 1992.

Walter, Ingo. 1993. "The Battle of the Systems: Control of Enterprises and the Global Economy." *Kieler Vorträge* No 122, Institut für Weltwirtschaft an der Universität, Kiel.

World Bank. 1996. World Development Report. *From Plan to Market*. Washington, DC: World Bank.

Chapter 5

Ahuja, Vinod and Deon Filmer. 1996. "Educational Attainment in Developing Countries: New Estimates and Projections Disaggregated by Gender." Journal of Educational Planning and Administration 3: 229–54.

Ahuja, Vinod, Benu Bidani, Francisco Ferreira and Michael Walton. 1997. "Everyone's Miracle? Revisiting Poverty and Inequality in East Asia." Directions in Development. The World Bank. Washington, DC.

Aoki, Masahiko, Hyung-Ki Kim, and Masahiro Okuno-Fujiwara, eds. 1996. *The Role of Government in East Asian Economic Development: Comparative Institutional Analysis*. New York: Oxford University Press.

Birdsall, Nancy and Richard H. Sabot. 1993. "Virtuous Circles: Human Capital Growth and Equity in East Asia." Background paper for *East Asian Miracle*. World Bank, Washington, D.C.

Blejer, Mario I. and Isabel Guerrero. 1991. "The Impact of Macroeconomic Policies on Income Distribution: An Empirical Study of the Philippines." *Review of Economics and Statistics*.

Deininger, Klaus and Lyn Squire. 1996. "A New Data Set Measuring Income Inequality." *The World Bank Economic Review* 10 (3): 565–91.

Diwan, Ishac and Michael Walton. 1997. "How International Exchange, Technology, and Institutions Affect Workers: An Introduction." *The World Bank Economic Review* 11(1): 1–15.

Ferreira, Francisco H. G. and Julie A. Litchfield. Forthcoming. "Education or Inflation? The Roles of Structural Factors and Macroeconomic Instability in Explaining Brazilian Inequality in the 1980s." LSE-STICERD Discussion Paper. London School of Economics.

Filmer, Deon and Lant Pritchett. 1997. "Child Mortality and Public Spending on Health." Policy Research Working Paper 1864. Development Research Group. World Bank, Washington, D.C.

Hammer, Jeffrey S., James Cercone and Ijaz Nabi. 1995. "Distributional Effects of Social Sector Expenditures in Malaysia, 1974 to 1989." In Dominique van de Walle and Kimberly Nead, eds., *Public Spending and the Poor: Theory and Evidence*. Baltimore, Md.: The Johns Hopkins University Press.

Jalan, Jyotsna and Martin Ravallion. 1997. "Spatial Poverty Traps?" Policy Research Working Paper 1862. Development Research Group. World Bank, Washington, D.C.

Kim, Dae-Il and Robert H. Topel. 1995. "Labor Markets and Economic Growth: Lessons from Korea's Industrialization, 1970–1990." In Richard B. Freeman and Lawrence F. Katz, eds., *Differences and Changes in Wage Structures*. Chicago: University of Chicago Press.

Pencavel, John H. 1995. "The Role of Labor Unions in Fostering Economic Development." Policy Research Working Paper 1469. World Bank, Washington, D.C.

Radhakrishna, R. and K. Subbarao. 1997. *India's Public Distribution System: A National and International Perspective*. World Bank Discussion Paper 380. World Bank, Washington, D.C.

Ranis, Gustav. 1995. "Another Look at the East Asian Miracle." *The World Bank Economic Review (International)* 9:509–34.

Ravallion, Martin. 1998. "Appraising Workfare Programs." Development Research Group. World Bank, Washington, D.C.

Ravallion, Martin and Shaohua Chen. 1997. "A Note on the Measurement of Income Inequality in Post-Reform Rural

China." Development Research Group. World Bank, Washington, D.C. Processed.

———. 1998. "Poverty Rates in East Asia with Zero Growth." Mimeo.

Subbarao, K., et al. 1997. *Safety Net Programs and Poverty Reduction: Lessons from Cross-Country Experience.* Directions in Development Series. World Bank, Washington, D.C.

Teranishi, Juro. 1996. "Sectoral Resource Transfer, Conflict and Macrostability in Economic Development: A Comparative Analysis." In Aoki et al., 1996.

UNESCO. 1997. UNESCO Statistical Yearbook. Paris: UNESCO Publishing.

United Nations. 1995. *World Urbanization Prospects: The 1994 Revision.* United Nations, New York, New York.

van de Walle, Dominique. 1995. "The Distribution of Subsidies through Public Health Services in Indonesia, 1978–87." In Dominique van de Walle and Kimberly Nead, eds., *Public Spending and the Poor: Theory and Evidence.* Baltimore, Md.: The Johns Hopkins University Press.

van de Walle, Dominique and Kimberly Nead. 1995. *Public Spending and the Poor: Theory and Evidence.* Baltimore, Md.: The Johns Hopkins University Press.

World Bank. 1990. *World Development Report 1990: Poverty.* New York: Oxford University Press.

———. 1993a. *The East Asian Miracle: Economic Growth and Public Policy.* New York: Oxford University Press.

———. 1993b. "Indonesia: Public Expenditures, Prices and the Poor." Report 11293-IND. Washington, D.C.

———. 1994. *Averting the Old Age Crisis: Policies to Protect the Old and Promote Growth.* World Bank Policy Research Report. Oxford University Press for the World Bank. Washington, D.C.

———. 1995. *World Development Report 1995: Workers in an Integrating World.* New York: Oxford University Press.

———. 1996a. *Involving Workers in East Asia's Growth.* Regional Perspectives on World Development Report 1995. World Bank, Washington, D.C.

———. 1996b. *World Development Report 1996: From Plan to Market.* New York: Oxford University Press.

———. 1997a. *China 2020: Development Challenges in the New Century.* China 2020 Series. Washington, D.C.

———. 1997b. *Sharing Rising Incomes: Disparities in China.* China 2020 Series. Washington, D.C.

Chapter 6

Cruz, W. and Repetto, R. 1992. *The Environmental Effects of Stabilization and Structural Adjustment Programs: The Philippines Case.* World Resources Institute; Washington, DC.

McMorran, R. T. and Hamilton, K. 1996. *Philippines: Scope for Integrating Macroeconomics and the Environment— Some Suggestions.* Draft; International Monetary Fund and the World Bank. Washington, DC.

World Bank. 1992. *Strategy for Forest Sector Development in Asia,* World Bank Technical Paper No. 182; World Bank; Washington, DC.

World Bank 1997. *Can the Environment Wait? Priorities for East Asia.* World Bank; Washington, DC.

WRI. 1998. *World Resources: A Guide to the Global Environment.* World Resources Institute; Washington, DC.

Chapter 7

International Monetary Fund. 1998. *World Economic Outlook.* May.

Systemic Bank Restructuring and Macroeconomic Policy. William Alexander et al., eds., Washington, DC: International Monetary Fund, 1997.

Robb, Caroline. 1998. *Social Impacts of the East Asian Crisis: Perceptions from Poor Communities.* Paper prepared for the East Asian Crisis Workshop, July 1998. Institute of Development Studies, University of Sussex, UK.

World Bank. 1996. *Managing Capital Flows in East Asia.* Washington, DC.

World Bank. 1997. *China 2020 Development Challenges in the New Century.* Washington, DC.